Beyond Innovation

Other Palgrave Pivot titles

Dirk Jacob Wolfson: The Political Economy of Sustainable Development: Valuation, Distribution, Governance

Twyla J. Hill: Family Caregiving in Aging Populations

Alexander M. Stoner and Andony Melathopoulos: Freedom in the Anthropocene: Twentieth Century Helplessness in the Face of Climate Change

Christine J. Hong: Identity, Youth, and Gender in the Korean American Christian Church

Cenap Çakmak and Murat Ustaoğlu: Post-Conflict Syrian State and Nation Building: Economic and Political Development

Richard J. Arend: Wicked Entrepreneurship: Defining the Basics of Entreponerology

Rubén Arcos and Randolph H. Pherson (editors): Intelligence Communication in the Digital Era: Transforming Security, Defence and Business

Jane Chapman, Dan Ellin and Adam Sherif: Comics, the Holocaust and Hiroshima

AKM Ahsan Ullah, Mallik Akram Hossain and Kazi Maruful Islam: Migration and Worker Fatalities Abroad

Debra Reddin van Tuyll, Nancy McKenzie Dupont and Joseph R. Hayden: Journalism in the Fallen Confederacy

Michael Gardiner: Time, Action and the Scottish Independence Referendum

Tom Bristow: The Anthropocene Lyric: An Affective Geography of Poetry, Person, Place

Shepard Masocha: Asylum Seekers, Social Work and Racism

Michael Huxley: The Dancer's World, 1920–1945: Modern Dancers and Their Practices Reconsidered

Michael Longo and Philomena Murray: Europe's Legitimacy Crisis: From Causes to Solutions

Mark Lauchs, Andy Bain and Peter Bell: Outlaw Motorcycle Gangs: A Theoretical Perspective

Majid Yar: Crime and the Imaginary of Disaster: Post-Apocalyptic Fictions and the Crisis of Social Order

Sharon Hayes and Samantha Jeffries: Romantic Terrorism: An Auto-Ethnography of Domestic Violence, Victimization and Survival

Gideon Maas and Paul Jones: Systemic Entrepreneurship: Contemporary Issues and Case Studies

Surja Datta and Neil Oschlag-Michael: Understanding and Managing IT Outsourcing: A Partnership Approach

Keiichi Kubota and Hitoshi Takehara: Reform and Price Discovery at the Tokyo Stock Exchange: From 1990 to 2012 palgrave▸pivot

Beyond Innovation: Technology, Institution and Change as Categories for Social Analysis

Thomas Kaiserfeld
Lund University, Sweden

palgrave
macmillan

© Thomas Kaiserfeld 2015

All rights reserved. No reproduction, copy or transmission of this publication may be made without written permission.

No portion of this publication may be reproduced, copied or transmitted save with written permission or in accordance with the provisions of the Copyright, Designs and Patents Act 1988, or under the terms of any licence permitting limited copying issued by the Copyright Licensing Agency, Saffron House, 6–10 Kirby Street, London EC1N 8TS.

Any person who does any unauthorized act in relation to this publication may be liable to criminal prosecution and civil claims for damages.

The author has asserted his right to be identified as the author of this work in accordance with the Copyright, Designs and Patents Act 1988.

First published 2015 by
PALGRAVE MACMILLAN

Palgrave Macmillan in the UK is an imprint of Macmillan Publishers Limited, registered in England, company number 785998, of Houndmills, Basingstoke, Hampshire RG21 6XS.

Palgrave Macmillan in the US is a division of St Martin's Press LLC, 175 Fifth Avenue, New York, NY 10010.

Palgrave Macmillan is the global academic imprint of the above companies and has companies and representatives throughout the world.

Palgrave® and Macmillan® are registered trademarks in the United States, the United Kingdom, Europe and other countries.

ISBN: 978–1–137–54711–8   EPUB
ISBN: 978–1–137–54712–5   PDF
ISBN: 978–1–137–54710–1   Hardback

This book is printed on paper suitable for recycling and made from fully managed and sustained forest sources. Logging, pulping and manufacturing processes are expected to conform to the environmental regulations of the country of origin.

A catalogue record for this book is available from the British Library.

A catalog record for this book is available from the Library of Congress.

www.palgrave.com/pivot

DOI: 10.1057/9781137547125

Contents

Preface		vi
1	Innovation Monomania	1
2	Technology, Institution and Change	11
3	Market Institutions	27
4	Evolutionary Economics	37
5	Performativity	41
6	Knowledge	47
7	Agency	57
8	Clusters, Systems and Blocks	67
9	Resistance to Change	77
10	Commons	88
11	Technological Determinism	93
12	Modernity and Its Critics	102
13	Postmodernity	111
14	Hybridity and Technology Transfer	119
15	Conclusions	128
Bibliography		135
Index		159

Preface

The background for this book is the increasing specialization of scholarly investigations of technological and institutional change. Although many studies of this particularly dynamic area are interdisciplinary, drawing on philosophy, history, sociology, economics and so on for methods, analytical frameworks and other necessary prerequisites, certain aspects have come to dominate over others. For instance, much of the research on technical change deals with innovation in firms and how to support it. The purpose of this book is to counterweigh this dominating firm-centric research trend by broadening our thinking about technological and institutional change. It is done by a multidisciplinary review of the most common ideas about the dynamics between technology and institutions. This book thus balances between being, on one hand, an introduction for students interested in different views and perspectives on technological and institutional change without paying attention to borders between different academic disciplines and, on the other, an instrument for cross-fertilization between different strands of thought and modes of analyses.

The ideas behind this text evolved in an interdisciplinary setting, the Sector Committee of Technology, Institution and Change, set up by the Swedish research foundation Riksbankens Jubileumsfond. The purpose of this group has been to discuss and initiate social and humanities research in the intersection between technical applications or political usages and economy, democracy and ethics. The underlying assumption is that elusive technological change in complex institutional environments indicates worthwhile research problems as well as research opportunities. There is simply a demand for updated knowledge about different social processes leading to institutional and technological dynamics.

One important strategy to achieve this is to review and bring together existing results from a variety of disciplines and research fields and this is what this book sets out to do. In addition, there is also a demand for a broader, critical and public discussion about which technologies to develop, by what means and with what ends. Insights from different experiences are necessary to develop this theme in order to pursue a democratic discourse. This book is thus written with the assumption that ideas materialized in texts actually can make a difference.

I want to express my appreciation, not only to Riksbankens Jubileumsfond, which set up the Sector Committee of Technology, Institution and Change making this endeavour possible, but also to the individual members of the committee who have all whole heartedly supported my attempts to bring the many different ideas and concepts together in one short book. Göran Blomqvist, managing director at Riksbankens Jubileumsfond, Mats G. Hansson, biomedical ethics at Uppsala University, Kristina Höök, human–machine interaction at KTH Royal Institute of Technology, Stockholm, Ericka Johnson, technology and social change at Linköping University, Fredrik Lundmark, research manager at Riksbankens Jubileumsfond, Cecilia Magnusson-Sjöberg, cyber law at Stockholm University, Christian Munthe, practical philosophy at University of Gothenburg, Lars J. Nilsson, environmental and energy systems studies at Lund University, Urban Strandberg, political science at University of Gothenburg, Jane Summerton, technology and social change at Linköping University and Nina Wormbs, history of technology at KTH Royal Institute of Technology, Stockholm, have all commented on earlier versions of this text. Thank you for your patience! I have also received many constructive and helpful comments from the seminar group in the Department of History of Science and Ideas at Uppsala University. In addition, anonymous reviewers have supplied very valuable comments improving the text considerably. I am alone responsible for all remaining shortcomings and errors.

palgrave▸pivot

www.palgrave.com/pivot

1
Innovation Monomania

Abstract: *Innovation has become the buzzword in a number of policy areas including research policy, economic policy and environmental policy. In the struggle against economic stagnation, innovation policies, sometimes in alliance with the academic field of innovation studies, promote dreams about institutions and technologies in which change can never be turned into nightmares. The ambition of this book is, however, to point to a number of alternative models and theories within the social sciences that describe or explain dynamics between institutions and technologies. The purpose is to demonstrate the rich multitude of ideas about technology, institution and change beyond innovation in the context of liberal markets.*

Keywords: innovation; innovation paradigm; innovation society; innovation studies

Kaiserfeld, Thomas. *Beyond Innovation: Technology, Institution and Change as Categories for Social Analysis*. Basingstoke: Palgrave Macmillan, 2015. DOI: 10.1057/9781137547125.0003.

Today's Western society is obsessed with change and more specifically technological change. Innovation has become the buzzword in a number of policy areas including research policy, economic policy and environmental policy.[1] There does not seem to be a single challenge to humanity that cannot be remedied by some technological innovation including more drastic methods such as geoengineering. This trust in technology is often enough combined with an equally strong trust in economic institutions such as liberal markets, an ideal type where suppliers of identical products compete for consumers with a minimum of regulations to govern behaviour. Definitions of innovation accordingly include notions of some sort of check like a demand for success in one way or another. A typical definition thus frames innovation as "ideas *successfully* applied in organizational outcomes and processes."[2] Consequently, considerable amounts of resources are funnelled into the development of different technologies or knowledge that is assumed to have a potential for commercialization.[3] For instance, the European Union framework programme for research and innovation running from 2014 to 2020 and named Horizon 2020 includes an initiative called Innovation Union. This is a "strategy to create an innovation-friendly environment that makes it easier for great ideas to be turned into products and services that will bring our economy growth and jobs."[4]

Innovation Union is a very representative symbol of the presently dominant view on technology, institution and change. With the right investments and the right strategy to create "an innovation-friendly environment", commercial success stories are supposed to follow to everyone's advantage. Such ideas rest on basic propositions of economic growth theory, "that in order to sustain a positive growth rate of output per capita in the long run, there must be continual advances in technological knowledge in the form of new goods, new markets, or new processes."[5] Important also are Joseph Schumpeter's notions of quality-improving innovations, as the engine of capitalist growth with entrepreneurs driven by a strong will as the cornerstone. Since new innovations destroy the results of earlier ones, making them obsolete, an important result is so-called creative destruction.[6] In the context of innovation, technological development is measured in commercial value, institutions according to how much they support the creation of new innovations, and, finally, change is always and in general beneficial. It is hardly surprising that innovation studies have thrived as an academic discipline in the social sciences and at business schools during the past decades.

Concepts such as "innovation society" and even "innovation paradigm" have been used to characterize the innovation monomania that dominates different policy initiatives, not the least in the European Union.[7] As a consequence, representatives of innovation studies boast about the appreciation of their programme.

> Innovation is increasingly recognized as a vitally important social and economic phenomenon worthy of serious research study. Firms are concerned about their innovation ability, particularly relative to their competitors, because they believe their future may depend on it. Politicians care about innovation, too, because of its presumed importance for growth, welfare, and employment. However, to recognize that innovation is desirable because of its assumed beneficial effects is not sufficient in itself. What is required is systematic and reliable knowledge about how best to influence innovation and exploit its effects to the full. Gaining such knowledge is the aim of innovation studies.[8]

Simultaneously, however, criticism against the academic and political focus on innovation as the high road to a better society has gained momentum.[9] This has also led to reconsiderations of different more or less implicit assumptions traditionally made within innovation studies. One such assumption is that institutions can be arranged in order to efficiently support innovation activities or innovation transfer and, as a consequence, that regions and countries with the best institutions for innovation get a competitive advantage when attracting investments.[10]

Admittedly, there are emerging sub-disciplines of innovation studies in which alternative aims are being formulated such as responsible innovation or sustainable innovation.[11] Under these headings, efforts are made by researchers from different backgrounds to broaden the field of innovation studies to include innovations that may not exclusively meet demands played out on some sort of market or be measured against their contribution to economic growth. Complementary ideals for innovation activities, they claim, should or could also contain appreciated qualities such as sustainability or fair trade. These variations are launched either accepting market success as the dominant prerequisite of innovations or, in the more radical version, denying it. Regardless of its different forms, innovation studies in general tries to form alliances with innovation policies to promote dreams about institutions and technologies in which change can never be turned into nightmares.

This is a review intended to moderate that dream. The ambition is to point to a number of alternative models and theories within the social sciences that describe or explain dynamics between institutions and technologies. The purpose is to demonstrate the rich multitude of ideas about technology, institution and change beyond innovation in the context of liberal markets. "Economics of innovation", as some critics have claimed, "does not, by and large, open the black box of technology, and it fails to engage with the increasingly sophisticated analyses of technology coming from history and sociology of technology".[12] Others have added, "there is in [innovation studies] little evidence of openness to debate on fundamental issues (assumptions, approaches, models, concepts, typologies, biases and limitations)".[13]

A case in point could be the problem of how to manage spent nuclear fuel from nuclear power plants, as of 2015 in operation in 31 different countries of the world. Already, there exists about 270,000 tonnes of spent nuclear fuel with an annual addition of about 12,000 tonnes of which 3,000 goes to reprocessing.[14] In order to develop methods to securely store this waste for the ten to hundred millennia it takes for the radioactivity to wear off, countries with nuclear power have different research and development programmes, many of them aiming for deep geological storage. The problems are to some extent technical, for instance, how to encapsulate the waste in order to seal it off from the environment in the storage. In addition, there are a number of social challenges, most importantly perhaps to develop democratic processes to decide the sites selected for storage.[15] While technical and scientific research for the development of methods for deep geological storage of spent nuclear fuel has been pursued since the 1970s, the development of social institutions for democratic site-selection processes is still an under-developed field despite many different efforts.

There are, to be sure, alternative solutions to the problem too, at least hypothetically. Today's conventional reactor technologies extract only small parts of the potential energy content of fissile material used in nuclear power plants. In order to improve efficiency, proponents of nuclear power want to develop so-called breeder reactors, which use the fuel more efficiently. Breeder technologies exist in different versions and stages of research and development which are sometimes named Generation IV reactors, or Gen IV for short. In common, however, they share a number of shortcomings making it likely to take decades of further research to develop them to be suitable for commercially competitive power production. But the research and development needed is uncertain in terms of outcomes while simultaneously being extremely expensive. The uncertainty surrounding Gen IV reactors has created a genuine insecurity regarding what to do with the spent nuclear fuel that has already been generated at existing nuclear power plants. Is the best solution really to store it in deep geological deposits more or less irretrievably or should this extremely radioactive and dangerous material be stockpiled for possible future use in Gen IV reactors although this may attract interest among those who want to access and use it to harm people?

From the traditional innovation perspective, the solution is of course to funnel resources to develop new reactor technologies while keeping the already spent nuclear fuel accessible. This is already going on in different national projects, in Europe through the European Atomic Energy Community (abbreviated to Euratom) and internationally in Generation IV International Forum with a number of partner countries as well as Euratom representing the European Commission and the member states of the European Union. In Europe, different initiatives to develop three different Generation IV reactor prototypes have been calculated to cost close to 11 billion euros supplied by the European Commission with additional support from different participating countries.[16]

In comparison, only negligible resources are spent on the problem of developing social institutions to secure necessary transports of spent nuclear fuel from power plants to reprocessing plants and back again. In addition, equally small resources are spent on the problem of how to set up institutions to secure the storage of spent nuclear fuel, either retrievably or irretrievably. Different social institutions are beginning to be reviewed and considered in order to find new ways of preserving and maintaining knowledge and memory regarding different sites and deposits for spent nuclear fuel. But this work is only starting and is not being pursued in any more systematic way. The problem is often a lack of knowledge about the multitude of existing institutions and their respective efficiency when set out to solve different problems.

This book is intended to supply a comprehensive review of different ways of thinking about how to combine changing technologies and institutions in order to meet different challenges, ideas that go beyond what is usually offered in innovation studies. This is in line with the present trend of more and more policymakers, researchers, journalists and laypersons realizing how our challenges are not only technical and scientific, but also social. As the European Commissioner for Research, Innovation and Science put it in the fall of 2013:

It takes profound knowledge and insight to really understand [...] challenges and how they affect us, and to guide us to solutions. That is why the Social Sciences and Humanities are more essential than ever, and why we, as policymakers, are keen to have their contribution. We need them to understand ourselves, our society and the challenges we face. We need them to guide politicians and policy makers and to inform public opinion. Research and technology provide many answers to the challenges we face, but technological fixes alone aren't enough to solve our major, complex problems. A knowledge society needs to know itself, and the social sciences and humanities are the keys to this.[17]

In order to develop sustainability, environmental innovations are not enough. In order to improve global health, medical research will have to be complemented, for instance, by new institutions for generic drugs. In order to avoid escalations of armed conflicts in different parts of the world, new weapons, for instance, unmanned drones, have seldom proved to be a long-lasting solution. Instead, although never a panacea, institutional change, for example, freedom from colonial oppression or a more balanced distribution of wealth and access to education, has time and again been demonstrated as radically more promising.

In order to engage humanities and social sciences in the global challenges of today, new interdisciplinary efforts are being mobilized under headings such as integrated humanities.[18] In the long run, it will not do to study technological change from the narrow perspective of innovations developed in singular institutional settings of market exchange. To meet the present challenges and be prepared for future ones, there is a need for broader theoretical approaches to the relations between technology and institution. Thus the demands for a broader review of theoretical frameworks already developed. Hopefully, it can also be used as a foundation for creating new ones so urgently needed at a time when financial crises have so far failed to generate new solutions beyond austerity measures or Keynesian expansion.[19]

The models and theories reviewed here have their origin in a number of disciplines such as history, anthropology, economics, political science, science and technology studies and so on, making this an interdisciplinary effort. Nevertheless, it is far from complete and the cases abound where classical texts have been referred to rather than updated lines of argument or more recently developed syntheses unless when something qualitatively new has been added. Thus, this review is not to be considered a research text aimed for the initiated and updated, but rather a thematically structured, yet incomplete survey intended for those interested in the multitude of existing perspectives on institution, technology and change. The aim is not to cover the wide range of theories of institution, technology and change in different specific areas, for instance, work organization or built environment, but to give samples of their variation.

With this short introductory discussion on innovation and the need for a complementary review, it is time to sketch its structure. Since the purpose of the book is to reveal some lines of thought in the lush vegetation of theories about technology, institution and change in order to supply alternatives to the contemporary focus on innovation as the salvation of our civilization, the second chapter will contain some definitions of the concepts of technology, institution and change. The purpose here is to define contours and shapes of the object of study as well as to discuss these concepts. In order to form a point of departure for the discussion of alternatives, the chapters following immediately after will deal with some of the institutional perspectives traditionally accepted in innovation studies. First, there will be some remarks on neoclassical economic theory including the concepts of invention and innovation in Chapter 3. The reason is that commercialization on a market traditionally has been seen as the ultimate success test within the academic field of innovation studies. The fourth chapter will deal with evolutionary economics. This is a theoretical approach that has used the mechanism of selection usually attributed to markets in order to understand processes on the supply side, another feature salient in innovation studies.

This is followed by a review in Chapter 5, of the more recent debate about performativity, that is, the potential applicability of innovation studies, which, from a policy perspective, is one of its most attractive features. Also, the present political focus on innovation and interest in innovation studies will be highlighted here, especially the tensions implicit in, on one hand, testing innovative success on liberal markets presumably out of political control where a myriad of consumers vote by individual choice and, on the other, the intentional planning and designing of institutions to make them support innovation work. The following chapters will then outline perspectives and salient features used when formulating theories about changes occurring in the relations between technology and institution. For instance, different forms of knowledge are often viewed as important both as generators of change and consequences of transformations of institutions and technologies. Likewise, the study of agency supplies another important perspective on changes of institutions and technologies.

The sixth chapter will thus deal with one of the most important components for the creation and development of commercially successful inventions and innovations according to economic growth theory and innovation studies, knowledge. Following this, agency in theories of technological and institutional change will be reviewed in Chapter 7 aligning with the most central ideas in innovation studies where the entrepreneur is still a cheered factor. Here, however, agency is presented broadly in order to complement the most important agent in innovation studies, the entrepreneur. After that follows Chapter 8 on clusters and systems, two of the more commonly mentioned institutional frameworks in innovation studies.

After this, in the nineth chapter, the review will start paying increasing attention to frameworks of analyses that are complementary to the perspectives usually supplied by innovation studies, first resistance to change followed by different forms of technological commons in Chapter 10. The eleventh chapter deals with another less developed theme in innovation studies, namely, determinism, which still, although to some extent dated, can shed light on the more or less eternal question about what is possible as opposed to what is unavoidable. Determinism is indeed a good introduction to the theories of technology and institution framed in Chapter 12 departing from the concepts of modernity. This is followed by a chapter on postmodernity. The last chapter is devoted to hybridity and ideas of technological and institutional transfer. Finally, the different perspectives are brought together in a concluding Chapter 15.

Notes

1. Lew Perren & Jonathan Sapsed, "Innovation as Politics: The Rise and Reshaping of Innovation in UK Parliamentary Discourse 1960–2005", *Research Policy* 42:10 (2013), 1815–1828.
2. Mark Dodgson & David Gann, *Innovation: A Very Short Introduction* (Oxford: Oxford University Press, 2010), 14, italics added.
3. See, for example, Jerry Courvisanos, *Cycles, Crises and Innovation: Path to Sustainable Development – A Kaleckian-Schumpeterian Synthesis* (Cheltenham: Edward Elgar Publishing, 2012), 40–62; Benoît Godin, "'Innovation Studies':

The Invention of a Speciality", *Minerva: A Review of Science, Learning and Policy* 50:4 (2012), 397–421.
4 Innovation Union homepage, accessed at: http://ec.europa.eu/research/innovation-union/index_en.cfm, March 17, 2014.
5 Philip Aghion & Peter Howitt, *Endogenous Growth Theory* (Cambridge, Mass: The MIT Press, 1998), 11.
6 Mark Elam, *Innovation as the Craft of Combination: Perspectives on Technology and Economy in the Spirit of Schumpeter*, Linköping Studies in Arts and Science 95 (Linköping: Linköping University, 1993); Philip Aghion & Peter Howitt, *The Economics of Growth* (Cambridge, Mass: The MIT Press, 2012), 85–101.
7 Sven Widmalm, "History of Science in the Age of Policy", in: *Aurora Torealis: Studies in the History of Science and Ideas in Honor of Tore Frängsmyr*, eds., Marco Beretta, Karl Grandin & Svante Lindqvist (Sagamore Beach: Science History Publications, 2008), 259–275; Sven Widmalm, "Innovation and Control: Performative Research Policy in Sweden", in: *Transformations in Research, Higher Education and the Academic Market: The Breakdown of Scientific Thought*, eds., Sharon Rider, Ylva Hasselberg & Alexandra Waluszewski, Higher Education Dynamics 39 (Dordrecht: Springer, 2013), 39–51.
8 Jan Fagerberg, Ben R. Martin & Esben S. Andersen, "Innovation Studies: Towards a New Agenda", in: *Innovation Studies: Evolution and Future Challenges*, eds., Jan Fagerberg, Ben R. Martin & Esben S. Andersen (Oxford: Oxford University Press, 2013), 1–17, p. 1.
9 Regarding the discussion on the research focus of technology studies, see: David Edgerton, "Innovation, Technology, or History? What Is the Historiography of Technology About?", *Technology and Culture* 51:3 (2010), 680–697.
10 Charlotte A. Cottrill, Everett M. Rogers & Tamsy Mills, "Co-citation Analysis of the Scientific Literature of Innovation Research Traditions: Diffusion of Innovations and Technology Transfer", *Science Communication* 11:2 (1989), 181–208.
11 For example, see: David H. Guston et al., "Responsible Innovation: Motivations for a New Journal", *Journal of Responsible Innovation* 1:1 (2014), 1–8.
12 Trevor Pinch & Richard Swedberg, "Introduction", in: *Living in a Material World: Economic Sociology Meets Science and Technology Studies*, eds., Trevor Pinch & Richard Swedberg (Cambridge, Mass: The MIT Press, 2008), 1–26, p. 1.
13 Benoît Godin, "'Innovation Studies': Staking the Claim for a New Disciplinary 'Tribe'", *Minerva* 52:4 (2014), 489–495, p. 493.
14 World Nuclear Association, accessed at: http://www.world-nuclear.org/info/nuclear-fuel-cycle/nuclear-wastes/radioactive-waste-management/, January 17, 2015.
15 Göran Sundqvist, *The Bedrock of Opinion: Science, Technology and Society in the Siting of High-Level Nuclear Waste* (Dordrecht: Kluwer Academic Publishers, 2002).

16 World Nuclear Association.
17 Quoted from: Paul Holm, Arne Jarrick & Dominic Scott, *Humanities World Report 2015* (Basingstoke: Palgrave Macmillan 2015), 174.
18 For example, see: Edward Slingerland, *What Science Offers the Humanities: Integrating Body and Culture* (Cambridge: Cambridge University Press, 2008).
19 Pierre Dardot & Christian Laval, *The New Way of the World: On Neoliberal Society*, original in French 2009 (London: Verso, 2013); Andrew Smithers, *The Road to Recovery: How and Why Economic Policy Must Change* (Chichester: John Wiley & Sons, 2013); David Harvey, *Seventeen Contradictions and the End of Capitalism* (London: Profile Books, 2014).

Further reading

Dodgson, Mark & David Gann (2010), *Innovation: A Very Short Introduction* (Oxford: Oxford University Press).

Fagerberg, Jan & Bart Verspagen (2009), "Innovation Studies – The Emerging Structure of a New Scientific Field", *Research Policy* 38, 218–233.

Fagerberg, Jan, Ben R. Martin & Esben S. Andersen, eds. (2013), *Innovation Studies: Evolution and Future Challenges* (Oxford: Oxford University Press, 2013).

Godin, Benoît (2012), "'Innovation Studies': The Invention of a Speciality", *Minerva: A Review of Science, Learning and Policy* 50:4, 397–421.

2
Technology, Institution and Change

Abstract: *In this chapter the key concepts of technology, institution and change are discussed and defined. The intimate relations between technology and institution are also described with the conclusion that technological and institutional change cannot be discussed in isolation from each other. Instead, they are different sides of the same coin, a coin with a value for an understanding of civilizations, global growth and cultural exchange as well as environmental sustainability, health and wealth distribution. Some concepts to be reviewed have been developed to frame the convergence of technology and institution in the sense that material expressions and social order are co-constructed or co-produced.*

Keywords: change; conservative invention; disruptive invention; incremental invention; institution; institutional lag; radical invention; technology

Kaiserfeld, Thomas. *Beyond Innovation: Technology, Institution and Change as Categories for Social Analysis*. Basingstoke: Palgrave Macmillan, 2015. DOI: 10.1057/9781137547125.0004.

It is nothing short of banal to claim that there are close relations between technology and institutions. The way we organize our lives to a large extent decides the way technologies are conceived and used while rules and regulations as well as silent agreements and practices are influenced, sometimes even determined, by hard-wired technologies. As a consequence, most analyses of institution and technology view the two entities as co-varying with each influencing the other in more or less equal measures. In much of more recent research, the relations between institutions and technology are further dissolved often resulting in a rejection of a division between the two altogether.[1]

But if we for a moment accept the existence of two different but partly overlapping categories named institution and technology, there are indications of their close relations everywhere we look. A classic car, for instance, has two seats in front and three in the back in order to fit most Western families. With a growing number of single households together with a rising awareness of the role of extensive car traffic play in urban air pollution, smaller cars are becoming more common. Still, there are of course larger alternatives for bigger families as well. This example shows how explicit and implicit agreements and disagreements we have regarding how to live and what to think put their imprints on different technical designs and technological solutions.

Vice versa, the fashioning of different technologies in construction and use also influence the mechanics of institutions.[2] Prior to the telephone, mail service was constant with mail deliveries several times a day in a typical Western city. With new communication technologies such as telegraph lines in the mid-19th century and a growing number of private telephones during the 20th century, mail service slowly transformed and was eventually more often reserved for important personal letters, official notes and, perhaps most common of all, bills. Today, only few people care to write letters at all, except maybe for Christmas. Official letters are to an increasing extent distributed in electronic formats, as are our bills. The ever-emptier letterboxes are proof of a transformed institution in turn demanding new regulations regarding the way personal communication and information are managed.[3]

Dated are nevertheless ideas of a general institutional lag meaning that institutions simulate technological activities.[4] Take, for example, US counterterrorism strategies involving Unmanned Aerial Vehicles (UAV), or drones, to accomplish targeted extra territorial and extra judicial killing operations. Here, specific characteristics of standard international law practices – boundaries, combatant status and neutrality laws – have been considered to address a legal framework for this technology and its use, not to mention how the approval, order and realization of operations have created a system involving a number of US intelligence organizations.[5] This is of course a reaction to the technically induced possibility of using UAVs and could be understood as institutional lag. But taking a step back, it is also clear that the technology of drones has been developed to accomplish extra territorial killings in an institutional setting where national borders on land and sea need to be recognized as well as national citizenship. In this institutional context, drones have proved the means justifiable by a military goal to thwart terrorist attacks in the United States.[6]

As a consequence, the common wisdom is presently that technologies and institutions influence one another. It is, however, harder to pinpoint conditions when such mutual stimulus appears.

> Technological change certainly can have independent effects on social structures. But technological change can have these effects only where the social organization exists that makes technology relevant. So for example, a hunter-gatherer society has little use for highways and telecommunications. But a society in which lowering the costs of transportation and communication makes it easier for firms to move goods and services to where there are opportunities to sell them gives huge incentives to firms that can figure out how to lower the costs.
>
> The creation of new technology is often viewed as resulting from the scientific manipulation of reality. The discovery of new technologies is often led by the perception that a solution to a particular problem would yield large monetary gains. But this is only part of what technology is. Technologies involve figuring out how to make goods and services that can be delivered, that are reliable, and that someone can be convinced to buy at prices at which the product can be produced.[7]

Thus, technological and institutional change is never discussed in isolation. Instead, it is to a growing extent seen as different sides of the same coin, a coin with a value for an understanding of civilization, global growth and cultural exchange as well as environmental sustainability, health and wealth distribution.[8] The perhaps best-known example of all, of how intricately related technological and institutional change is, stems from the influential idea of division of labour. Traditionally attributed to an example of a pin factory given by Adam Smith in his *Wealth of Nations* from 1776, the practice of making production more efficient through the definition and distribution of different phases in a production process on different work stations where individuals pursue their tasks in simultaneous coordination was well known long before that.[9] Already in 1701, clock manufacture using division of labour had been described in detail.

> Watch is a work of great variety, and 'tis possible for one Artist to make all the several Parts, and at last to join them altogether; but if the Demand of Watches shou'd become so very great as to find constant imployment for as many Persons as there are Parts in a Watch, if to every one shall be assign'd his proper and constant work, if one shall have nothing else to make but Cases, another Weels, another Pins, another Screws, and several others their proper Parts; and lastly, if it shall be the constant and only imployment of one to join these several Parts together, this Man must needs be more skilful and expeditious in the composition of these several Parts, than the same Man cou'd be if he were also to be imploy'd in the Manufacture of all these Parts. And so the Maker of the Pins, or Wheels, or Screws, or other Parts, must needs be more perfect and expeditious at his proper work, if he shall have nothing else to pusle and confound his skill, than if he is also to be imploy'd in all the variety of a Watch.[10]

The origin of division of labour is unknown, but has been attributed to among many others, the Swedish 17th-century mechanic Christopher Polhem.[11] Important here is, however, not so much the historical origin of this form of organization of work as the fact that division of labour becomes more efficient if the parts used are interchangeable since work requires less time if every part put together in a work station does not need to be specifically adjusted to fit. As a result, division of labour in mechanical assembly to a large extent relies on precision in the production of parts and has as a consequence that different artefacts can more easily be repaired if one of the parts fails. In this case, as in many others, institutions such as organization of work are intricately related to the available technologies such as machines and tools.

Some concepts have been developed to frame the convergence of technology and institution in the sense that material expressions and social order are co-constructed or co-produced.[12] One important work-related example of how technology and institution has been converging into co-construction is the development of numerically controlled machine tools. Scholars have pointed out how this technology of automation in the workplace was to a lesser extent the result of a strive for making production cheaper than for deskilling labour in order for machine-shop owners to be in control also of production that had earlier required skilled labour.[13] In this case, the institutional problem of controlling skilled workers found a technological solution with far-reaching consequences for the institution of work. Neither the control of skilled labour in workshops nor numerically controlled machine tools can be understood in isolation but are co-constructed.

From the same perspective of co-construction, environmental concerns define what is seen as relevant environment research just as environment research defines if policies should aim for restoration of nature or limit human exploitation of it. For instance, political concern about the problem of making communities more sustainable due to a reluctance to change human behaviour has induced a scientific interest in environmental humanities.[14] Vice versa, scientific hypotheses that human activity has changed the earth on a scale comparable with some major events of its past has led to the introduction of the informal geological term of anthropocene, which has raised political interest accordingly.[15] Insights that knowledge in different disciplines and areas of research, from the natural and medical to the social and humanistic, may be relevant in a variety of ways have resulted in calls for synthesis research across disciplinary boundaries.[16]

It is impressions of a contemporary and simultaneous technological and institutional revolution that will be the point of departure here with the aim to eventually discuss how today's technological and institutional contexts may underpin a view on the relations between institutional and technological change that corresponds to the present global context rather than a struggle between nations for economic growth through technological innovation.[17] This reasoning relies on the hypothesis that theories regarding the dynamics of technology and institution most often depend on the current state of these categories and perhaps even more so on imaginaries about their future.[18] But first, a short discussion of technology and institution as well as change is needed.

Definitions of technology abound and range from the very narrow idea of technology being machines to a very wide notion encompassing virtually everything man-made, material or immaterial. Irrespective of which definition is chosen, it is clear that the concept of technology developed in tandem with industrialization and slowly replaced the concept of art when referring to knowledge about machines, mechanics, production processes or the application of different sciences such as chemistry or mechanics.[19] In the vernacular, technology is often seen as representing a specific sort of materiality connected to industrial products or crafts as well as material structures and systems. In other contexts, technology can also be viewed as any tool, material or immaterial, that can be used for a specific end or purpose.

Regardless of which definition of technology one embraces, it should be clear that most of them include aspects of specific types of knowledge and practices. The knowledge referred to often deals with the understanding of how different phenomena can be controlled and used in purposeful ways. It may also include analytical skills regarding processes of production or logistics. The practices that may be connected to the context of technology are often skills that can only seldom be formally presented, but are more often practical in the sense that they include movements and the use of the senses, skills that can be acquired only through experience. In fact, these different defining components of technology mirror a struggle within the engineering community about the nature of technological knowledge. Some claim it is to be viewed as a form of science, but where the object of study is artificial rather than natural. Others prefer to highlight alternative components of technology where tacit knowledge and the need for drawings and models as well as the way to conceive innovations in the "mind's eye" are crucial.[20] When narrowing the concept further, materiality is often brought into play as well as connections to methods of production.

In general, however, if the syllabi of engineering programmes were to stake out the area of knowledge particularly suited for different technical practices, there is no question that mathematics and other natural sciences play a pivotal role in technology. In addition, there are often important parts of engineering programmes devoted to organizational skills as well as knowledge of the social and cultural dynamics of processes important for the ability to make a group of engineers, a company or even representatives for different parts of society, to agree on, or at least form a majority in favour of, a proposal for a specific technical solution, whether the stretch of a new railroad track or standards for a district heating system.

As living conditions transform, so do the opinions regarding the appropriate skill of an engineer. And if it is reasonable to assume that engineers are the social carriers of technology, this implies that the content of the concept of technology is in constant flux.[21] Without having made any systematic analyses of the content of engineering programmes and on-going discussions regarding their content, the trend today seems to be towards larger shares of social science and humanities or at least a discussion regarding the advantages of integrating such perspectives.

But engineering science and other areas judged necessary for engineers to be trained in is only one aspect of the content of technology, a perspective often connected to what is seen as important for the ability to innovate. Just as important is, however, the knowledge needed in order to operate or use existing technologies. There is a growing interest among researchers regarding this aspect of technological practices and furthermore how users are participating in the development and design of technologies. Users developing software for personal computers and other electronic devices, for instance, in the form of mobile applications software, or apps, for smartphones, is an example of this phenomenon.[22]

Taken together, technology is a nebulous concept with varying meanings in different contexts. Classically, it is often defined by phrases such as the systematic and organized use of combinations of knowledge and materials to achieve assumed foreseeable results for human purposes.[23] Departing from this definition and the ambiguities inherent in terms such as "systematic and organized" it is important to point out that it does not contradict the perspective of critical theory as described by Andrew Feenberg:

> Critical theory argues that technology is not a thing in the ordinary sense of the term, but an "ambivalent" process of development suspended between different possibilities. This "ambivalence" of technology is distinguished from neutrality by the role it attributes to social values in the design, and not merely the use, of technical systems. On this view, technology is not a destiny but a scene of struggle. It is a social battlefield, or perhaps a better metaphor would be a *parliament of things* on which civilizational alternatives are debated and decided.[24]

What Feenberg here manages to include are the two main perspectives that can be used to analyse technology, its use or functional value and its exchange or symbolic value. An expensive sports car is seldom used to its full capacity, especially not in contexts where there are legal and social institutions limiting speed. Its symbolic value is thus higher than its functional value.

Needless to say, Feenberg's is an extremely wide definition with the advantage that it delineates technology as a contested category while holding most doors open. Since the very definition of technology is one of many tools that can be used to sort technologies into different categories such as worthwhile and worthless, this one fits well with the aim of this review. As already mentioned, the concept of innovation stresses commercial value and technological development as a way to achieve economic growth. It can be contrasted to other forms of technology, for instance, systems developed to recycle and redistribute artefacts in demand such as eyeglasses.[25] Such practices span another alternative in which commercial value is less important. The point here is that both alternatives can be called technologies, albeit in different institutional contexts, and their respective value is an issue to be debated and discussed in the parliament of things.

Social institutions are generally thought of as relatively stable social patterns of human activity organized to manage fundamental problems such as producing life-sustaining resources, reproducing individuals and sustaining social structures in a given environment.[26] The much wider view is also sometimes represented, that institutions are simply social norms or conventions. The key point is that individuals are expected to conform to these institutional norms and conventions. For instance, Douglass North has defined institutions by claiming they "are the rules of the game in a society, or more formally, are the humanely devised constraints that shape human interaction".[27] A more specific definition of institutions, while maintaining generality, has been formulated:

> [Institutions seen] *as building-blocks of social order*: [...] represent socially sanctioned, that is, collectively enforced expectations with respect to the behavior or specific categories of actors or to the performance of certain activities. Typically, they involve *mutually related rights and obligations* for actors, distinguishing between appropriate and inappropriate, "right" and "wrong," "possible" and "impossible" actions and thereby organizing behaviour into predictable and reliable patterns.[28]

In both these cases of definitions of institutions, technology is viewed as exogenous, that is, as a factor external to the category. To continue the metaphorical definition of North, institutions are the rules of the game while technology partly constitutes the constraints underlying the game just as basic human needs for survival do. Simultaneously, the building blocks of social order constitute the institutions while technology is an exogenous force defining the building blocks and mechanisms for putting them together. It is nevertheless this second definition of institutions that will be used here.

Institutions can be both formal – regulated by a code that can be changed only through the application of some predetermined processes – and informal in which the pattern remains stable through silent agreement or consent. Formal institutions are the national systems of criminal courts while Christmas celebrations to a large extent constitute an informal institution in today's Western society, although institutions of different Christian congregations may keep some of its formal aspects. Most institutions actually have both a formal and an informal aspect where the informal patterns are sometimes described as institutional cultures. It is also common that institutions overlap and intersect as the example of Christmas celebrations show.

But considering the wide definition of technology introduced earlier, there are strong conceptual links between the categories of technology and institution. If technology is a scene of struggle, a parliament of things, institutions are involved in collectively enforcing the dos and don'ts of technology. Conversely, institutions rely on knowledge about material conditions among many other things and are thus dependent on technologies. In fact, technologies such as digital platforms for social networking may simultaneously be institutions just as institutions often involve large measures of technology in order to exist. One example of such a perspective is the more specific so-called rise of network society since the 1970s including organizations, technologies and meaning.[29] This is a theme often represented in the literature on modernism and postmodernism. More examples will be given in the chapters devoted to that discourse.

So just as there are marked differences between the two categories of institutions and technologies, there are intimate ties between them. For instance, institutions are often decisive for the establishment of both symbolic and exchange values as well as functional and use values of a specific technology. Likewise, technologies are as important for the establishment of values of a specific institution.

The close interaction between technology and institution has also been framed as a balancing act between change and inhibition. Technology as a combination of different tools for problem-solving activities is continuous, cumulative and dynamic in turn leading to accelerated progress. The process is counterweighed by institutions understood as ceremonial practices relying on traditional hierarchy and status. To some extent, such institutions may inhibit technological change.[30] The classic car mentioned in the beginning of this chapter represents this perceived dichotomy between technological change and institutional inhibition. From this perspective, the car simultaneously mirrors both technological change in terms of more efficient transport and institutional conservation by confirming and regenerating the established Western institution of a family with two parents and two or three children who can be fitted in the back.[31]

The close and complex interrelations have consequences in the sense that changes in one category is likely to have immediate and unavoidable consequences for the other and vice versa. Or, alternatively, changes in one category will lead to adaptation in the other for continued functionality. Change, then, is a classical philosophical problem, from the antiquities, when Heraclitus claimed that no man ever steps in the same river twice, at least according to a formulation by Plato, a statement opposed by the Eleatics, and onwards. In this context, it is enough to conclude that there are different forms of change, from the genuinely originally new, which in some settings may lead to the evolution of altogether new institutions, to amendments and editions of already developed technologies within an established institutional context.

In Judaeo-Christian tradition, change relying on human creativity, whether originally new or just minor alterations, has often been viewed as problematic, even heretical. Since God created the world out of nothing, creatio ex nihilo, any attempt to similarly create inventions out of nothing was considered conceit.[32] With modernity, such notions were more seldom expressed. Instead, our culture seems to be obsessed with the understanding of change, institutional and technological as well as cultural and social. Accordingly, analyses of change have become a major raison d'être for much of social and human research.

It is in this context the concepts of invention and innovation have come to dominate research on technology, institution and change. If invention is thought of as any type of technical novelty and innovation is an invention that has arrived as a product on a market, change in terms of new technologies is often framed through the concepts of invention and innovation, which together span an imagined timeline from idea to market. It is important, however, to remember that such a timeline is in no way deterministic. It is possible that inventions may be abundant in a specific culture without ever being developed into innovations. One often cited example is ancient China where a lot of techniques such as paper, gunpowder, printing and so on was invented, but never implemented on a broader scale.[33] Thus, it is possible to have inventions in abundance and still lack innovations. Inventions are nevertheless a necessary precondition for innovations for which an imagined market is the ultimate test of success.[34]

Inventions generating altogether new institutions are sometimes called radical (or disruptive) in the sense that they in hindsight have proved to lead to radically new possibilities – breakthroughs that shatter the way of doing things.[35] They can be contrasted to conservative (or incremental) inventions, which are adjustments and improvements to existing technologies. One example of a radical invention could be the development of early devices for the generation, transmission and utility of electricity such as generators and electrical motors that exploit phenomena of electromagnetism.[36] A potentially radical invention of the future could be a device for wireless power transfer.[37] Another category of inventions is sometimes referred to as transformational to be seen as even more revolutionary than radical inventions. They not only change the nature of processes, services or products, but also affect whole economies. Similarly, the concept of socio-technical transitions have been developed to frame system changes with far-reaching social repercussions.[38] In reality, time is a crucial factor since an invention at first judged incremental may after years or decades be valued as radical and after another half-century or so as transformational.

The key difference between epistemic ruptures and incremental changes of technology is that accumulated knowledge plays a more important role in the latter case than in the former.[39] If technology is viewed as changing incrementally, institutions become more important for the accumulation of knowledge and practices than in the case of radical change. Vice versa, in cases of radical technological change, institutions should be designed to question existing knowledge and practices if change is to be promoted. Note also that technologies and institutions are likely to co-vary so that institutions supporting cumulative knowledge and practices simultaneously underpin incremental technological change rather than radical and vice versa.

In terms of institutional change, there are corresponding distinctions made. Periods of social transformation, for instance, have often been seen as unsettled times interrupting settled ones where social relations are more stable. The characteristics of these different types of transformation have been attributed to the differences in the ways of which culture supply habits and skills from which action is assembled where of course potential cultural components are new technologies.[40] Others have claimed that institutional structures dominate in settled times, whereas unsettled times are more influenced by agency.[41] Of course, it is also possible to discuss institutional change in terms of invention and innovation, perhaps implying more conscious planning of rules and regulations as well as informal restrictions and encouragement of behaviours. Concepts such as social innovation and social entrepreneurship have been developed to frame strategies and activities aimed to meet social needs such as wealth distribution, alternatively new ideas about social relations, organizations and institutions.[42]

Simultaneously, however, these concepts can be viewed as just another way of expanding the innovation paradigm including the notion that any problem can be solved through the application of scientific methods and planning. Or as a strategy has been succinctly formulated by a number of predominantly Austrian and German social scientists in the so-called Vienna declaration of 2011, it is to embed "the concept of social innovation in a comprehensive theory of innovation".[43] This entails broadening the concept of innovation beyond the notion of a new technology successful on some sort of market. Social innovations are thus not primarily technological and they may be measured by acceptance in other institutional settings than market contexts.[44]

Still, market settings are the most common and most important institutional environment in innovation studies. This will therefore be discussed in more detail in the next chapter although when reviewing different theoretical approaches to social innovation and potential resources, the scope is admittedly broader.

Notes

1 Bruno Latour & Steve Woolgar, *Laboratory Life: The Construction of Scientific Facts* (London: SAGE Publications, 1979).
2 For a famous example, see: Langdon Winner, "Do Artifacts Have Politics?", *Daedalus: Journal of the American Academy of Arts and Sciences* 109:1 (Winter 1980), 121–136; Bernward Joerges, "Do Politics Have Artefacts?", *Social Studies of Science* 29:3 (1999), 411–431; Steve Woolgar & Geoff Cooper, "Do Artefacts Have Ambivalence: Moses' Bridges, Winner's Bridges and Other Urban Legends in S&TS", *Social Studies of Science* 29:3 (1999), 433–449; Bernward Joerges, "Scams Cannot Be Busted: Reply to Woolgar & Cooper", *Social Studies of Science* 29:3 (1999), 450–457.
3 Claude S. Fischer, *America Calling: A Social History of the Telephone to 1940* (Berkeley, Calif: University of California Press, 1992).
4 Clarence E. Ayres, *The Theory of Economic Progress* (Chapel Hill: The University of North Carolina Press, 1944), 170–187. See also: David Hamilton,

"Ayres' *Theory of Economic Progress*: An Evaluation of Its Place in Economic Literature", *American Journal of Economics and Sociology* 40:4 (1981), 427–438.
5   Michael Coleman & David H. Gray, "The Legality of Targeted Killings and the Use of Drones in the War on Terror", *Global Security Studies* 5:1 (2014), 37–55.
6   Milena Sterio, "The United States' Use of Drones in the War on Terror: The (Il)legality of Targeted Killings under International Law", *Case Western Reserve Journal of International Law* 45:1–2 (2012), 197–214.
7   Neil Fligstein, *The Architecture of Markets: The Economic Sociology of Twenty-First-Century Capitalist Societies* (Princeton: Princeton University Press, 2001), 4.
8   Norbert Elias, "Technization and Civilization", *Theory, Culture & Society* 12:3 (1995), 7–42.
9   Adam Smith, *An Inquiry into the Nature and Causes of the Wealth of Nations*, 2 vols. (London, 1776).
10 Henry Martyn, *Considerations upon the East-India Trade* (London, 1701), 43–44, See also: Andrea Maneschi, "The Tercentenary of Henry Martyn's *Considerations upon the East-India Trade*", *Journal of the History of Economic Thought* 24:2 (2002), 233–249.
11 Ken Alder, *Engineering the Revolution: Arms and Enlightenment in France, 1763–1815* (Princeton: Princeton University Press, 1997), 221.
12 Peter Taylor, "Co-construction and Process: A Response to Sismondo's Classification of Constructivisms", *Social Studies of Science* 25:2 (1995), 348–359; Sheila Jasanoff, "The Idiom of Co-production", in: *States of Knowledge: The Co-production of Science and Social Order*, ed., Sheila Jasanoff (London: Routledge, 2004), 1–12.
13 Harry Braverman, *Labor and Monopoly Capital: The Degradation of Work in the Twentieth Century* (New York: Monthly Review Press, 1974); David F. Noble, *Forces of Production: A Social History of Industrial Automation* (New York: Random House, 1984). Karl Marx presented similar ideas, see: Langdon Winner, *Autonomous Technology: Technics-Out-of-Control as a Theme in Political Thought* (Cambridge, Mass: The MIT Press, 1977), 38–40.
14 Sverker Sörlin, "Environmental Humanities: Why Should Biologists Interested in the Environment Take the Humanities Seriously?", *BioScience* 62:9 (2012), 788–789.
15 Jan Zalasiewicz et al., "The New World of the Anthropocene", *Environmental Science & Technology* 44:7 (2010), 2228–2231.
16 Stephen R. Carpenter et al., "Accelerate Synthesis in Ecology and Environmental Sciences", *BioScience* 59:8 (2009), 699–701.
17 Gregory Tassey, *The Technology Imperative* (Cheltenham: Edward Elgar Publishing, 2007).
18 See: Wolfgang Streeck, "Institutions in History: Bringing Capitalism Back In", Max Planck Institute for the Study of Societies, Discussion Paper 09/8, November 2009.

19　David F. Noble, *America by Design: Science, Technology, and the Rise of Corporate Capitalism* (Oxford: Oxford University Press, 1977); Eric Schatzberg, "Technik Comes to America: Changing Meanings of Technology in America before 1930", *Technology and Culture* 47:3 (2006), 488–512; Leo Marx, "Technology: The Emergence of a Hazardous Concept", *Technology and Culture* 51:3 (2010), 561–577; Eric Schatzberg, "From Art to Applied Science", *Isis* 103:3 (2012), 555–563.

20　Edwin T. Layton, *The Revolt of the Engineers: Social Responsibility and the American Engineering Profession* (Cleveland: Press of Case Western Reserve University, 1971); Walter G. Vincenti, *What Engineers Know and How They Know It: Analytical Studies from Aeronautical History* (Baltimore: Johns Hopkins University Press, 1990); Eugen S. Ferguson, *Engineering and the Mind's Eye* (Cambridge, Mass: The MIT Press, 1992).

21　Charles Edquist & Olle Edqvist, "Social Carriers of Techniques for Development", *Journal of Peace Research* 16:4 (1979), 313–331.

22　Nelly Oudshoorn & Trevor Pinch, eds., *How Users Matter: The Co-construction of Users and Technology* (Cambridge, Mass: The MIT Press, 2005).

23　Edwin T. Layton, Jr., "Technology as Knowledge", *Technology and Culture* 15:1 (1974), 31–41.

24　Andrew Feenberg, *Critical Theory of Technology* (Oxford: Oxford University Press, 1991), 14.

25　Jacqueline Ramke, Renee Du Toit & Garry Brian, "An Assessment of Recycled Spectacles Donated to a Developing Country", *Clinical & Experimental Ophthalmology* 34:7 (2006), 671–676; John Szetu et al., "Recycled Donated Spectacles: Experiences of Eye Care Personnel in the Pacific", *Clinical & Experimental Ophthalmology* 35:4 (2007), 391–392.

26　Seumas Miller, "Social Institutions", in: *The Stanford Encyclopedia of Philosophy* (Fall 2012 Edition), ed., Edward N. Zalta, accessed at: http://plato.stanford.edu/archives/fall2012/entries/social-institutions/, January 17, 2014.

27　Douglass North, *Institutions, Institutional Change and Economic Performance* (Cambridge: Cambridge University Press, 1990), 3. Here quoted from: Christopher Kingston & Gonzalo Caballero, "Comparing Theories of Institutional Change", *Journal of Institutional Economics* 5:2 (2009), 151–180, p. 154.

28　Wolfgang Streeck & Kathleen Thelen, "Introduction: Institutional Change in Advanced Political Economies", in: *Beyond Continuity: Institutional Change in Advanced Political Economies*, eds., Wolfgang Streeck & Kathleen Thelen (Oxford: Oxford University Press, 2005), 3–39, p. 9.

29　Manuel Castells, *The Information Age: Economy, Society and Culture*, 3 vols. (Oxford: Blackwell, 1996–1998).

30　Clarence E. Ayres, *The Theory of Economic Progress*; Warren J. Samuels, "Technology Vis-à-Vis Institutions in the JEI: A Suggested Interpretation", *Journal of Economic Issues* 11:4 (1977), 871–895.

31 For a different view of cars as efficient personal transport, see: Ivan Illich, *Energy and Equity*, World Perspectives 49 (New York: Harper & Row, 1974), 15–19.
32 Wladyslaw Tatarkiewicz, *A History of Six Ideas: An Essay in Aesthetics*, Melbourne International Philosophy Series 5 (The Hague: Martinus Nijhoff Publishers, 1980); David N. Perkins, "The Possibility of Invention", in: *The Nature of Creativity: Contemporary Psychological Perspectives*, ed., R. J. Sternberg (Cambridge: Cambridge University Press, 1988), 362–385.
33 Patrick K. O'Brien, "The Needham Question Updated: A Historiographical Survey and Elaboration", *History of Technology* 29 (2009), 7–28.
34 William J. Baumol, *The Free-Market Innovation Machine: Analyzing the Growth Miracle of Capitalism* (Princeton: Princeton University Press, 2002), 9–10.
35 For a number of different meanings of the concept, see: Stephen G. Green, Mark B. Gavin & Lynda Aiman-Smith, "Assessing a Multidimensional Measure of Radical Technological Innovation", *IEEE Transactions on Engineering Management* 42:3 (1995), 203–214; Elicia Maine & Elizabeth Garnsey, "Commercializing Generic Technology: The Case of Advanced Materials Ventures", *Research Policy* 35:3 (2006), 375–393.
36 Herbert W. Meyer, *A History of Electricity and Magnetism*, Burndy Library Publication No. 27 (Norwalk, Conn: Burndy Library, 1972); Thomas P. Hughes, *Networks of Power: Electrification of Western Society, 1880–1930* (Baltimore: Johns Hopkins University Press, 1983).
37 "Wireless Power", *Wikipedia*, accessed at: http://en.wikipedia.org/wiki/Wireless_power#See_also, April 9, 2014.
38 Jochen Markard, Rob Raven & Bernhard Truffer, "Sustainability Transitions: An Emerging Field of Research and Its Prospects", *Research Policy* 41:6 (2012), 955–967.
39 Edward W. Constant II, *The Origins of the Turbojet Revolution* (Baltimore: Johns Hopkins University Press, 1980).
40 Ann Swidler, "Culture in Action: Symbols and Strategies", *American Sociological Review* 51:2 (1986), 273–286.
41 Ira Katznelson, "Periodization and Preferences: Reflections on Purposive Action in Comparative Historical Social Science", in: *Comparative Historical Analysis in the Social Sciences*, eds., James Mahoney & Dietrich Rueschemeyerd (Cambridge: Cambridge University Press, 2003), 270–303.
42 Michael D. Mumford, "Social Innovation: Ten Cases from Benjamin Franklin", *Creativity Research Journal* 14:2 (2002), 253–266.
43 Hans-Werner Franz, Josef Hochgerner & Jürgen Howaldt, "Challenge Social Innovation: An Introduction", in: *Challenge Social Innovation*, eds., Hans-Werner Franz, Josef Hochgerner & Jürgen Howaldt (Dordrecht: Springer, 2012), 1–16, p. 4.
44 James A. Phills, Jr., Kriss Deiglmeier & Dale T. Miller, "Rediscovering Social Innovation", *Stanford Social Innovation Review* 6:4 (2008), 34–43; Geoff

Mulgan, "Social Innovation Theories: Can Theory Catch Up with Practice?", in: *Challenge Social Innovation*, eds., Hans-Werner Franz, Josef Hochgerner & Jürgen Howaldt (Dordrecht: Springer, 2012), 19–42.

Further reading

Franz, Hans-Werner, Josef Hochgerner & Jürgen Howaldt, eds (2012), *Challenge Social Innovation* (Dordrecht: Springer).

Perkins, David N. (1988), "The Possibility of Invention", in: *The Nature of Creativity: Contemporary Psychological Perspectives*, ed., R. J. Sternberg (Cambridge: Cambridge University Press), 362–385.

Marx, Leo (2010), "Technology: The Emergence of a Hazardous Concept", *Technology and Culture* 51:3, 561–577.

Schatzberg, Eric (2012), "From Art to Applied Science", *Isis* 103:3, 555–563.

3
Market Institutions

Abstract: *Different aspects of market institutions are reviewed. The idea of the unregulated market is an ideal construction and all real markets are regulated in some way or another. Indicative of markets is the exchange between a supplier, which is supposed to make a profit by supplying a product or a service to a consumer who is to compensate the supplier for this service. To that end, discussions on which forms of markets are suitable for the promotion of innovations are reviewed. A conclusion is that complicated institutional conditions influence the mechanics of markets. Another is that different recently developed concepts in general point towards a trend of closer and closer ties between producers, suppliers, consumers and users.*

Keywords: development pairs; dominant design; market architecture; mediation junction; neocorporatism

Kaiserfeld, Thomas. *Beyond Innovation: Technology, Institution and Change as Categories for Social Analysis.* Basingstoke: Palgrave Macmillan, 2015. DOI: 10.1057/9781137547125.0005.

When analysing technical change in terms of institutional conditions, the traditional and still most commonly used institution is different forms of markets where supply and demand are supposed to meet.

Once upon a time, many people believed that "The Market" was something that has always existed in a quasi-Natural state, much like gravity or language. It seemed to enjoy a material presence, sharing many of the characteristics of nature, and as such was deemed a coherent unified phenomenon.[1]

Indicative of markets is the exchange between a supplier, which is supposed to make a profit by supplying a product or a service to a consumer who is to compensate the supplier for this service. More generally, markets constitute a type of institutionalized exchange between more than two actors, often more than three so that competition exists. To that end, there are thorough discussions on which forms of markets are to be considered suitable for the promotion of innovations. Traditionally, markets where private companies compete for demand of products have been considered efficient.[2] More recently, however, public authorities have also started to be seen as important promoters of technical change through the use of procurement processes where technologies are contracted that not yet exist on the market. Such specific markets functioning as protected spaces where radical inventions and innovations can be developed secluded from open market competition is sometimes called niches.[3] Historically, it has mainly been military technologies such as weapons systems or communication systems that has been promoted by public procurement. But with the advent of the welfare state in the 20th century, other technologies such as medical and transport have also to a large extent relied on public procurement rather than private markets for development.[4] Also, educational technologies such as digital whiteboards have to a large extent been developed with public customers such as schools in mind.

Sometimes, the difference between private and public markets has been moderated while the dissimilarities have been underscored at others, for instance, by highlighting distinctions between coordinated and liberal markets expressed in different clusters of institutions.[5] One important conclusion developed as a consequence of the distinction between different forms of markets has been that the efficiency of economic institutions in promoting technical change to a large extent relies on how efficiency is measured. If, for instance, the measure is economic efficiency in the development of new technology and improvements of existing ones, price elasticity of demand has been pointed out as an important factor. Technologies that are in price-inelastic demand tend to be more efficiently developed on coordinated markets, for instance, through public procurement processes. Conversely, if demand is price elastic, liberal consumer markets tend to be judged more efficient in promoting technical change since process innovation, that is, innovation in the production process, is more important for lowering production costs than product innovation, that is, innovation in the products themselves.

When these two features are brought together, a recurring pattern of historical development can sometimes be observed. In the early stages of developing new technologies, investments are often high making the products expensive and the number of consumers limited to big organizations or states, simultaneously leading to price-inelasticity if demand is high enough among the few and resource-rich customers. If it is a case of emergence of new technology demanding altogether new markets, radical (or disruptive) innovations may lead to the existing competence being obsolete, in turn leading to disruptive market relations creating new links between producers and buyers.

In this early phase, innovation is dominated by push on the supply side, whether based on scientific knowledge or other forms of skills that still are out of demand. In this context, early adopters, a small number of often well-educated consumers inclined to risk buying a new product, are regularly considered an important segment.[6] After the establishment of innovation on a market, new products can be developed using either existing technologies in order to establish new market relations or disruptive technologies for already enrolled buyers. In the last phase, regular innovations are introduced on established markets.[7] Now, the pull on the demand side has become more important than the know-how developed on the supply side. In these situations, customers often want to influence design, and markets are often more efficient if they are coordinated. As more entities are sold the price becomes lower and the product will be standardized through some dominant design.[8] Simultaneously, a less coordinated market is supposed to become more efficient for the promotion of technical change although at this stage, more often in the production process than in the product itself.[9]

Obviously, diffusion of innovations to a large extent depends on market institutions, at least theoretically. Research on innovation diffusion has, for instance, shown how users on a new market tend to grow along an S-shaped curve with early adopters generally being well-educated with higher social status more frequently adopting innovations and thus forming an opinion leadership in turn persuading their peers.[10] Traditional innovation studies have consequentially paid a lot of interest to diffusion phenomena, for instance, mapping different institutional conditions that promote diffusion of innovation.

An example of this type of technical change is the computer. Originally, the computer was usually a big and heavy calculation machine bought only by large organizations in demand of calculation power or data storage capacity. As digital components became cheaper and smaller, shrinking computers could still supply the calculation power needed for assignments such as word processing or for simple computer games. When personal computers were developed as a dominant design in the late 1970s and early 1980s, they were put on more regular consumer markets through outlets such as RadioShack. Although stories abound on how highly priced products travel from coordinated markets to mass markets, a parallel example to the computer is the automobile some 20–30 years earlier; this is of course nothing predetermined. For each example of a technological product that has become cheap enough to buy and simple enough to operate for a large number of potential consumers, there are equal numbers of products that have not made the same trajectory despite high hopes among suppliers. Private jets or helicopters are still not sold off the shelf on regular markets.

This is just a small sample of the different types of existing markets, ranging from the ideal liberal market with suppliers selling similar products to an infinite number of consumers to the thoroughly institutionalized marketplace of reality with a limited number of consumers choosing between dissimilar products. In theory, there are of course many more factors that can be used to characterize different markets, describe their peculiar conditions for exchange of products and analyse their influence on technical change. Strictly speaking, markets are about the exchange of value, not function in terms of commodities or services.[11] In the context of invention and innovation, however, it is important to remember that commercialization on some type of a market is the ultimate litmus test of success. In the ideal world of innovation studies, it is supposedly here and nowhere else that the value of an innovation is finally and conclusively determined.[12]

A more elaborate example of how market conditions for technical change may play out could depart from the concept of mediation junction, defined as an institutional and organizational space where different stakeholders, typically potential users or consumers or their representatives, negotiate technological developments on arenas such as the state, the market and civil society.[13] Especially consumers and users have been increasingly engaged in product development in areas such as mobile apps. The successful negotiations between different intermediary actors representing diverse interests, for instance, regarding the rules and norms to regulate an exchange of some sort, is thought to rely on the existence of a recognized mediation junction. Functioning mediation junctions are thus a necessary, though not sufficient condition for the influence of intermediary actors on the establishment of some sort exchange regime.

The facilitation and creation of mediation junctions are an example of how market conditions can be created through market architecture, that is, the social construction, or more correctly the result of long historical evolution, of the extensive institutional support needed in order to supply the conditions necessary for the efficient exchange of goods and services. Inherent in the concept of market architecture is the idea that markets should indeed be understood as institutions and that they evolve and transform accordingly. Of course, regulatory bodies are important for market architecture and it has moreover been proved that stability is an important feature for keeping transaction costs down:

> In order to take advantage of new technology, firms need to establish stable relationships to their suppliers, workers, and principal competitors. The ability to establish these relationships is itself dependent on the production of stable societal institutions such as governments and law.[14]

Property rights stand out in this context. Developed in Western society in early modernity, they not only include patenting systems for inventions, but also make it possible to allocate resources by, for instance, using land assets as security for loans used to invest in new technologies. By stressing the priority of stability before even competitiveness and profits, market architecture to a large extent mirrors the wisdom of new institutionalism – to be described and surveyed in Chapter 9 – applied to a neo-classical world. This becomes even more evident as the boundaries of markets are stretched in order to also include education policies as well as employment patterns. A number of national alternatives surface: German cartels, French central state planning, Japanese state–industry collaboration and Scandinavian capital–labour cooperation.

The market concept can thus easily be broadened and many attempts have indeed been presented to generalize interaction between actors beyond transactions of goods and service, production and consumption. One such is the concept of action arenas where action situations are created for participants interacting according to exogenously set rules and conditions. The functioning of the action arena is evaluated according to evaluation criteria taking into account among other things the outcomes of the interactions.[15]

But markets are also the result of concrete technologies for transport and communication as well as technologies developed for trading and consumption such as mail order or cold store for frozen goods. In addition, market architecture can thus be understood as different actors' and actor groups' concrete efforts to format trading conditions. In this more narrow sense, market architecture is constituted by "the components that frame and underlie the actual exchange, and ensure the necessary stability, and information flow for the process. This includes the physical infrastructure, technologies needed for communication and facilitation of exchange, institutions and specific legislation and regulations on the market."[16] One instrument in this vein is efforts to make a specific product the industrial standard.[17]

Expanding the frame of reference, the existence of mediation junctions can be connected to the political organization of neocorporatism where social needs are generally satisfied by tripartite negotiations between public officials on behalf of politically elected decision makers, suppliers representing commercial industry and users often represented by consumer organizations such as patient interest groups in the case of medical technologies. Neocorporatism has often been connected with social democracy.[18] Nevertheless, the same type of tripartite negotiations between capital, labour or consumption and the state has been observed in many other political settings, especially in smaller European countries.[19]

Taken together, the model suggested here departs from a political setting where neocorporative solutions to public technological problems and needs lay the ground for institutional space – mediation junctions, to be used for negotiations between different intermediary actors – which together with traditions and ideologies form an exchange regime with isomorphic tendencies, that is, a trend to organize in similar ways, efforts with different aims. Of course, such a framework is a simplification. Exchange regimes are also constituted by cultural components as well as historical and social factors beyond negotiations and decision making.

Historian of technology, Mats Fridlund, has termed one type of mediation junction development pairs. They are defined as close, long-term collaboration between a private industrial company and a state customer on several development projects for new technologies.[20] Development pairs tend to evolve in specific political settings such as those marked by neocorporatism, but France also has been pointed out as a country where development pairs exist. Here engineers with similar education, often at the same engineering schools, negotiate as representatives for business and public authorities. The successful long-term and close collaboration implied by the existence of a specific development pair implies trust between the two parties and a potential for the state customer to have demands for innovative technology at reasonable price met while the private supplier is guaranteed a credible customer who will stand by agreements as well as a site of reference to use when putting the procured innovative technology on the market.

Mats Fridlund then goes on to show how similar patterns of long-term and close cooperation appear among development pairs in different sectors in one neocorporative setting. Developments of electro-technology, telephony, railway trains and so on in 20th-century Sweden can all be characterized as the result of development pairs. The reason is that in high-technology industries with thorough technical change, professionals tend to dominate over management, setting their own working procedures, thus facilitating cooperation with colleague professionals across legal and organizational borders.[21] Since 1995, however, Sweden has joined the European Union where development pairs are hard to maintain due to strict procurement regulations.

This example shows how concepts used to lend understanding to different parts of an institutional landscape, such as mediation junction and development pairs, can be combined into conceptual chains, ranging from the abstract and general to the more concrete and specific. Concepts can of course be constructed in infinitely many combinations in order to point out both similarities and differences. In this case, the reward has been the demonstration of the complicated institutional conditions that often may influence the mechanics of different markets. More specifically, the concepts reviewed here point towards the often described trend of closer and closer ties between producers, suppliers, consumers and users on different forms of markets.

DOI: 10.1057/9781137547125.0005

Notes

1. Philip Mirowski & Edward Nik-Khah, "Command Performance: Exploring What STS Thinks It Takes to Build a Market", in: *Living in a Material World: Economic Sociology Meets Science and Technology Studies*, eds, Trevor Pinch & Richard Swedberg (Cambridge, Mass: The MIT Press, 2008), 89–128, p. 89.
2. William J. Baumol, *The Free-Market Innovation Machine: Analyzing the Growth Miracle of Capitalism* (Princeton: Princeton University Press, 2002).
3. Jochen Markard, Rob Raven & Bernhard Truffer, "Sustainability Transitions: An Emerging Field of Research and Its Prospects", *Research Policy* 41:6 (2012), 955–967, p. 957.
4. Fred Block & Matthew R. Keller, eds, *State of Innovation: The U.S. Government's Role in Technology Development* (Boulder, Co: Paradigm Publishers, 2010); Mariana Mazzucato, *The Entrepreneurial State: Debunking Public vs. Private Sector Myths* (London: Anthem Press, 2013).
5. Peter A. Hall & David Soskice, "An Introduction to Varieties of Capitalism", in: *Varieties of Capitalism: The Institutional Foundations of Comparative Advantage*, eds, Peter A. Hall & David Soskice (Oxford: Oxford University Press, 2001), 1–68.
6. Everett M. Rogers, *Diffusion of Innovations* (New York: The Free Press, 1962).
7. Vivien Walsh, "Invention and Innovation in the Chemical Industry: Demand-Pull or Discovery-Push?", *Research Policy* 13:4 (1984), 211–234; William J. Abernathy & Kim B. Clark, "Innovation: Mapping the Winds of Creative Destruction", *Research Policy* 14:1 (1985), 3–22; Benoît Godin & Joseph P. Lane, "Pushes and Pulls: Hi(S)tory of the Demand Pull Model of Innovation", *Science, Technology, & Human Values* 38:5 (2013), 621–654.
8. Johann Peter Murmann & Koen Frenken, "Toward a Systematic Framework for Research on Dominant Designs, Technological Innovations, and Industrial Change", *Research Policy* 35:7 (2006), 925–952.
9. James M. Utterback & William J. Abernathy, "A Dynamic Model of Product and Process Innovation", *Omega* 3:6 (1975), 639–656; Atsushi Akiike, "Where Is Abernathy and Utterback Model?", *Annals of Business Administrative Science* 12 (2013), 225–236, accessed at: http://www.gbrc.jp/journal/abasjp/ms/abas12-17.pdf, March 18, 2014.
10. Everett M. Rogers, "Where Are We in Understanding the Diffusion of Innovations?", in: *Communication and Change: The Last Ten Years – and the Next*, eds, Wilbur Schramm & Daniel Lerner (Honolulu: The University Press of Hawaii, 1976), 204–222.

11 Arjun Appadurai, "Introduction: Commodities and the Politics of Value", in: *The Social Life of Things: Commodities in Cultural Perspective*, ed., Arjun Appadurai (Cambridge: Cambridge University Press, 1986), 3–63.
12 Benoît Godin, "'Innovation Studies': The Invention of a Speciality", *Minerva: A Review of Science, Learning and Policy* 50:4 (2012), 397–421.
13 Ruth Oldenziel, Adri A. Albert de la Bruhèze & Onno de Wit, "Europe's Mediation Junction: Technology and Consumer Society in the 20th Century", *History and Technology* 21:1 (2005), 107–139. See also Ruth Schwartz Cowan, "The Consumption Junction: A Proposal for Research Strategies in the Sociology of Technology", in: *The Social Construction of Technological Systems*, eds, Wiebe E. Bijker, Thomas P. Hughes & Trevor Pinch (Cambridge, Mass: The MIT Press, 1989), 261–280; Karin Zachmann, "A Socialist Consumption Junction: Debating the Mechanization of Housework in East Germany, 1956–1957", *Technology & Culture* 43:1 (2002), 73–99; Nelly Oudshoorn & Trevor Pinch, eds, *How Users Matter: The Co-construction of Users and Technology* (Cambridge, Mass: The MIT Press, 2005).
14 Neil Fligstein, *The Architecture of Markets: The Economic Sociology of Twenty-First-Century Capitalist Societies* (Princeton: Princeton University Press, 2001) 5.
15 Ellinor Ostrom, *Understanding Institutional Diversity* (Princeton: Princeton University Press, 2005).
16 Gabriel Söderberg, *Constructing Invisible Hands: Market Technocrats in Sweden 1880-2000*, Acta Universitatis Upsaliensis: Uppsala Studies in Economic History 98 (Uppsala: Uppsala University, 2013), 15. See also: Caitlin Zaloom, *Out of the Pits: Traders and Technology from Chicago to London* (Chicago: Chicago University Press, 2006).
17 Fligstein, *The Architecture of Markets*, 72.
18 A discussion on the Swedish welfare state and consumption can be found in Helena Mattsson & Sven-Olov Wallenstein, "Introduction", in: *Swedish Modernism: Architecture, Consumption and the Welfare State*, eds, Helena Mattsson & Sven-Olov Wallenstein (London: Black Dog Publishing, 2010), 6–33; Mattias Tydén & Urban Lundberg, "In Search of the Swedish Model: Contested Historiography", in: *Swedish Modernism*, 36–49.
19 Peter J. Katzenstein, *Small Countries in World Markets: Industrial Policy in Europe* (Ithaca, NY: Cornell University Press, 1985); Bo Rothstein, "State Structure and Variations in Corporatism: The Swedish Case", *Scandinavian Political Studies* 14:2 (1991), 149–171; Bo Rothstein, *Den korporativa staten: Intresseorganisationer och statsförvaltning i svensk politik* (Stockholm: Norstedts, 1992).
20 Mats Fridlund, *Den gemensamma utvecklingen: Staten, storföretagen och samarbetet kring den svenska elkrafttekniken* (Stockholm: Symposion, 1999).
21 Fligstein, *The Architecture of Markets*, 119.

Further reading

Fligstein, Neil (2001), *The Architecture of Markets: The Economic Sociology of Twenty-First-Century Capitalist Societies* (Princeton: Princeton University Press).

Hall, Peter A. & David Soskice, eds (2001), *Varieties of Capitalism: The Institutional Foundations of Comparative Advantage* (Oxford: Oxford University Press).

4
Evolutionary Economics

Abstract: *This chapter deals with the family of theories that use the metaphor of natural selection for free-market economy stipulated by neo-classic theory. Here, the inherent logic of technological change is illustrated by genetic variation whereas the mechanisms of decision made on a market as well as the institutions surrounding it correspond to the selection pressure exercised by the environment in natural selection. Change is thus generated by endogenous factors, and the evolutionary process takes place due to learning and imitation rather than replication. Another important consequence is that the focus of analyses is set on processes and institutions on the supply side rather than on the demand side, which is often the case in analyses of market dynamics.*

Keywords: evolutionary economics

Kaiserfeld, Thomas. *Beyond Innovation: Technology, Institution and Change as Categories for Social Analysis*. Basingstoke: Palgrave Macmillan, 2015. DOI: 10.1057/9781137547125.0006.

An often-used metaphor for the free-market economy stipulated by neo-classic theory is that of natural selection. Here, the inherent logic of technological change is illustrated by genetic variation whereas the mechanisms of decision made on a market as well as the institutions surrounding it correspond to the selection pressure exercised by the environment in natural selection.[1] This simple version of the metaphor can be compared to a situation in which technologies are continuously tried in an existing environment, and the one that on the whole is most efficient for the time being is adopted until the environment is changed to favour some other technology or new alternatives emerge that prove more efficient again.

Evolutionary economics is a collective term for a strand of institutional theories that further develops the metaphor of natural selection applied to markets. The central idea is to neglect individual behaviour and instead stress selection on different entities such as routines or social groups and in this way make technology theoretically endogenous. This means that technological change is brought into the theory rather than being an external force that influences institutions without itself being understood. Change is thus generated by endogenous factors, and the evolutionary process takes place due to learning and imitation rather than replication.[2] Another important consequence from the perspective of innovation studies is that the focus of analyses is set on processes and institutions on the supply side rather than on the demand side, which is often the case in analyses of market dynamics.

Evolutionary theories of economics depart from the notion that a company can be viewed as a phenotype that is fitted to the ever-changing economic environment although fitness here is defined as profitability. While the company corresponds to the phenotype, routines, defined as a collection of procedures together resulting in predictable and specifiable outcomes within the organization, corresponds to the genotype of a particular company.[3] In most evolutionary models, the company employs scientific methods and information as well as other means to make processes and products fit existing (market) conditions better.

From the 1950s, evolutionary theories have been developed as a response to neo-classical economic theories where market-driven demand has been regarded the most important generator of growth. In evolutionary theories, the supply side including technology is stressed as well. Here, investments in research and development constitute only one component. In addition, professional development in engineering and scientific training is highlighted together with the continuous improvements mirroring gained understanding of different processes involved. Moreover, interaction between processes and technologies is assumed to be strong leading to feedback between technical and institutional change. In more generalized terms, both technologies and institutions may supply constraints as well as possibilities and they co-evolve as long as entrepreneurs are learning how to mix the available technologies and institutions into productive routines, which in turn supply new technologies and institutions. Examples of co-evolution between technology and institutions, or physical and standardized social technologies as they have been referred to, abound:

> The car, the airplane, the transistor, the computer, and the laser all surfaced as new technologies of potentially wide applicability, but requiring considerable work and ingenuity before they would be worth anything in economic use. It took a long time, and a lot of investments, and much learning, and learning how to learn [...] before these new technologies became major contributors to economic growth. A common feature to the development paths taken by major new technologies is that quite unforeseen capabilities and users are discovered along the route. Different new technologies often interact in complex and surprising ways.[4]

Such insights into how technologies and institutions can be embraced as endogenous entities in one and the same theoretical framework relies on the observation of how economic growth has developed in different countries during different time periods, in the United States in the decades before and after the turn-of-the-century, 1900, in Japan during the 1960s and 1970s and in Korea, Taiwan, Singapore and Taiwan until the 1990s. Important in this context is the ability to not only copy existing technologies from abroad and introduce more efficient production methods through investments in better organized educational systems and restructuring of business life, but also support entrepreneurial initiatives and innovation.[5]

The simple metaphor of evolution can of course be developed further, and one of the more successful attempts has been innovation researcher Maureen McKelvey's effort to complicate matters by differentiating between science and technology as activities as well as government and market as institutional sources for seeking new knowledge.[6] In her co-evolutionary model, new knowledge drives the emergence of technological alternatives, while the selection among alternatives is a social process that does not necessarily lead to maximized profitability. Thus, scientific and economic environments have different criteria for what is to be considered incentives, success and so on.

Notes

1. Harvey Brooks, "Technology, Evolution and Purpose", *Dædalus* 109:1 (Winter 1980), 65–81.
2. Christopher Kingston & Gonzalo Caballero, "Comparing Theories of Institutional Change", *Journal of Institutional Economics* 5:2, 151–180.
3. Richard R. Nelson, "Recent Evolutionary Theorizing about Economic Change", *Journal of Economic Literature* 33:1 (March 1995), 48–90; Richard R. Nelson & Bhaven N. Sampat, "Making Sense of Institutions as a Factor Shaping Economic Performance", *Journal of Economic Behaviour & Organization* 44:1 (2001), 31–54.
4. Richard R. Nelson, *Technology, Institutions, and Economic Growth* (Cambridge, Mass: Harvard University Press, 2005), 30.
5. Richard R. Nelson & Howard Pack, "The Asian Miracle and Modern Growth Theory", *The Economic Journal* 109:457 (July 1999), 416–436.
6. Maureen McKelvey, *Evolutionary Innovations: The Business of Biotechnology* (Oxford: Oxford University Press, 1996).

Further reading

Nelson, Richard R. (2005), *Technology, Institutions, and Economic Growth* (Cambridge, Mass: Harvard University Press).

5
Performativity

Abstract: *Different perspectives on the intervention of social sciences in general and economic theories in particular as well as expertise in the formation of institutions, so called performativity, are discussed. In the field of institutional and technological change, the issue of scholarly work as opposed to activism has been discussed at length. The question is reflexive: whether different notions of technological and institutional change may contribute to a change of existing relations between technologies and institutions. There are of course important examples of this; one needs only to mention Marxism as both a theory and a political practice. The conclusion is nevertheless that technological and institutional change influences the forming of theories more than theories influencing policymaking and practices.*

Keywords: activism; economic performativity; representationalism; virtualism

Kaiserfeld, Thomas. *Beyond Innovation: Technology, Institution and Change as Categories for Social Analysis*. Basingstoke: Palgrave Macmillan, 2015. DOI: 10.1057/9781137547125.0007.

Another important idea connected to institutional preconditions of technological change is performativity. This is also the first notion highlighted in this review that is not at the core of innovation studies, although more often connected to the discipline of science and technology studies. At the heart of performativity is the notion that scientists, when constructing scientific theories about reality, also influence reality, not simply mirror it. In fact, this aspect has been an important part of science since early modernity and has often been used to highlight the potential utility of scientific knowledge.[1] More recent and aspiring claims, including institutional consequences, have been framed in the concept of technoscientific imaginaries broadly defined as "collectively imagined forms of social life and social order reflected in the design and fulfilment of nation-specific scientific and/or technological projects". For the sake of precision, it may also be worthwhile to point out that the demand on technoscientific imaginaries to be "nation-specific" has been dropped in later versions thus giving the concept a broader domain to operate on.[2]

In other words, there is no reason to limit influence on reality to natural scientists. It may equally well apply to the social sciences and the study of institutions. In social science, the concept of performativity has been developed to take on a number of different meanings, from informed activism and gender performativity over concepts such as speech acts to discursive practices.[3] At the heart is still, however, the capacity to intervene in the course of events. Performativity can thus be contrasted to representationalism, which denotes the ideal of standing back without interfering with the processes being studied, or at least to reduce meddling when possible.

In this context, performativity will be used to denote the intervention of social science in policymaking and especially policies aiming for the creation of institutions and conditions supporting inventions and innovations in order to boost economic growth. As can be expected, there are close connections between different forms of market architecture and performativity. Applicability in general, and policy implications in particular, has namely been one of the most important attractions of innovation studies outside of academia.[4] In this context, the process of disentanglement between the social and the ethical on one hand and the economic on the other is one very important aspect of performativity. This can be contrasted to the opposite but intertwined process of entanglement between producers and consumers in which wishes and hopes of both parties are to be calibrated.[5] None of these themes has escaped dispute.

In the debate about performativity, two main lines of argument have recently crystallized.[6] The first one stresses how (primarily economic) theories influence the design of existing institutions through expert advice.[7] When, for instance, trying to come to terms with undesired consequences of a certain technology such as carbon dioxide emissions from industrial processes such as steel manufacturing, in economic terms called a negative externality since it is a cost that affects someone who has not chosen to incur the emission, many different institutional solutions can be developed to frame this externality. One of these, the mimic of a neo-classical market where emission rights can be traded, has been realized in the European Union Emission Trading Scheme in operation since 2005.[8]

Another phenomenon often managed through the construction of a neo-classical institution is overflow, something that occurs when costs are purposely generated despite the value of benefits gained being lower. An example is when a fisher lands bigger catches than is motivated by the incomes he or she can hope to collect from it. There may be many different motifs behind such behaviour, for instance, symbolic value in large catches. In cases of externalities and overflows, institutions have been developed in order to keep control over the effects, often mimicking neo-classical markets.[9]

This stance, named economic performativity here, can be contrasted to another notion of performativity called "virtualism", the idea that economic thought shapes economic practice and its institutions in terms of markets and transactions through the abstraction of economic activity as well as of the consumer.[10] Economic models drive economic relations so that real activities and behaviour conform to the predicted ideals. The result is a realization of the virtual reality of economic thought. At first glance, the ideas of virtualism seem similar to those expressed by economic performativity. In reality, though, the differences are drastic since virtualism puts the ritual behaviours within the frame that defines the practices that are to be analysed whereas economic performativity frames the economic relations and interactions and how they are formed by economic theory as the object of analyses.[11] Virtualists claim that in their case, "what lies within the frame is *not* the market system as an actual practice, but on the contrary a ritualized expression of an ideology of the market".[12] If virtualists claim that expertise cultures decide real economic relations and transactions through the generation of economic models, economic performativity implies the mutual formation of models and relations.

The differences between the two perspectives on performativity presented here is that the view of economic performativity limits markets to a neo-classical price mechanism while virtualism brings in broader aspects of value exchange such as symbolic value, culture and so on.[13] From an institutional perspective, it would make sense to view performativity à la virtualism. A focus on innovation economy, however, often limits the understanding of performativity to neo-classical price mechanisms and, beyond them, the formulation of different policy implications. The role of social science itself, and the performative potential of broader notions of institutional and technological change, is less often reflected upon.

Apparently, ideas about performativity can thus be expanded to include institutional theory as well as any other theory about the conditions in the world. In the field of institutional and technological change, for instance, the issue of scholarly work as opposed to activism has been discussed at length.[14] The question here is reflexive: whether the different concepts and notions regarding technological and institutional change may in fact underpin changes in the existing relations between technologies and institutions. In general, there are of course important examples of occurrences where analyses have led to transformed relations. One needs only to mention Marxism as both a theory, which will be given more room in Chapter 11, and a political practice.

More interesting is, however, if the difference between economic performativity and virtualism is reflected in different views on whether theories of institutions and technologies frame regularities of the dynamic relations between them or if they are instead better viewed as a ritualized expression of practices belonging to an activist culture of social sciences. Such ideas touch upon the very foundations of social science, to represent and analyse or to influence and change, and are beyond the scope of this text. The general hypothesis here is that present relations between technology, institution and change are much more important in influencing the theories about them than the theories are in their influence on the relations.

There are nevertheless very strong tendencies in today's social science to engage more in different debated topics, not only through the triangulation of science, technology and social movements.[15] The ensuing debate on the future of humanities and social sciences and the emergence of a number of new interdisciplinary problem areas – often framed in terms such as medical humanities or digital humanities integrating the problem areas of humanities and social sciences with those of medical specialists or computer specialists and artists of different media – has led to a revival of the conviction that competence in social science, arts and humanities as well as science, technology and medicine is needed in order to handle different contemporary challenges, may they be environmental, technological or other.

These trends in today's social science have sometimes been brought together under the heading of interactive humanities. In common, they have an ambition to also bring humanities and social sciences into the framework of participatory science where the presentation of research results is expected to be paralleled by engagement in the use of the achieved results. Although relying on a centuries-long history, such ideas have acquired renewed momentum during the past decades.

Notes

1 Owen Hannaway, "Laboratory Design and the Aim of Science: Andreas Libavius versus Tycho Brahe", *Isis* 77:4 (1986), 585–610.
2 Sheila Jasanoff & Sang-Hyun Kim, "Containing the Atom: Sociotechnical Imaginaries and Nuclear Power in the United States and South Korea", *Minerva* 47:2 (2009), 119–146; Sheila Jasanoff & Sang-Hyun Kim, eds, *Dreamscapes of Modernity: Sociotechnical Imaginaries and the Fabrication of Power* (Chicago: The University of Chicago Press, forthcoming August 2015).
3 Judith Butler, *Bodies that Matter: On the Discursive Limits of Sex* (London: Routledge, 1993); Karen Badar, "Posthumanist Performativity: Toward an Understanding of How Matter Comes to Matter", *Signs: Journal of Women in Culture and Society* 28:3 (2003), 801–831.
4 Benoît Godin, "'Innovation Studies': The Invention of a Speciality", *Minerva: A Review of Science, Learning and Policy* 50:4 (2012), 397–421.
5 Alex Preda, "STS and Social Studies of Finance", in: *The Handbook of Science and Technology Studies: Third Edition*, eds, Edward J. Hackett et al. (Cambridge, Mass: The MIT Press, 2008), 901–920.
6 For a short review of different views on the relations between theory and reality, see: Patrik Aspers, "Theory, Reality, and Performativity in Markets", *The American Journal of Economics and Sociology* 66:2 (2007), 379–398.
7 Michel Callon, "What Does It Mean to Say That Economics Is Performative?", in: *Do Economists Make Markets? On the Performativity of*

  Economics, eds, Donald MacKenzie, Fabian Muniesa & Lucia Siu (Princeton: Princeton University Press, 2007), 311–357.
8 Larry Lohmann, "Carbon Trading, Climate Justice and the Production of Ignorance: Ten Examples", *Development* 51:3 (2008), 359–365; Larry Lohmann, "Neoliberalism and the Calculable World: The Rise of Carbon Trading", in: *The Rise and Fall of Neoliberalism: The Collapse of an Economic Order?*, eds, Kean Birch & Vlad Myhnenko (London: Zed Books, 2010), 77–93.
9 Michel Callon, "An Essay on Framing and Overflowing: Economic Externalities Revisited by Sociology", in: *Laws of the Markets*, ed., Michel Callon (Oxford: Blackwell, 1998), 244–269.
10 James G. Carrier & Daniel Miller, *Virtualism: A New Political Economy* (Oxford: Berg, 1998).
11 Daniel Miller, "Turning Callon the Right Way Up", *Economy and Society* 31:2 (2002), 218–233.
12 Ibid., 224. See also: Michel Callon, "Why Virtualism Paves the Way to Political Impotence", *Economic Sociology: European Electronic Newsletter* 6:2 (2005), 3–20, accessed at: http://econsoc.mpifg.de/archive/esfeb05.pdf, February 20, 2014; Daniel Miller, "Reply to Michel Callon", *Economic Sociology: European Electronic Newsletter* 6:3 (2005), 3–13, accessed at: http://econsoc.mpifg.de/archive/esjuly05.pdf, February 20, 2014.
13 Aspers, "Theory, Reality, and Performativity in Markets".
14 Vasilis Galis & Anders Hansson, "Partisan Scholarship in Technoscientific Controversies: Reflections on Research Experience", *Science as Culture* 21:3 (2012), 335–364.
15 David Hess, Steve Breyman, Nancy Campbell & Brian Martin, "Science, Technology, and Social Movements", in: *The Handbook of Science and Technology Studies: Third Edition*, eds, Edward J. Hackett et al. (Cambridge, Mass: The MIT Press, 2008), 473–498.

Further reading

Galis, Vasilis & Anders Hansson (2013), "Partisan Scholarship in Technoscientific Controversies: Reflections on Research Experience", *Science as Culture* 21:3, 335–364.

Pinch, Trevor & Richard Swedberg, eds (2008), *Living in a Material World: Economic Sociology Meets Science and Technology Studies* (Cambridge, Mass: The MIT Press).

6
Knowledge

Abstract: *From the 1950s, precursors to innovation theory developed tools to take into account factors such as research and the learning of new practices in order to understand technological change as a factor behind economic growth. Another theme in this chapter is the different views on how to combine knowledge in order to achieve institutional and technological change. Traditionally, knowledge has been viewed as an individual capacity or even trait. During more recent decades, however, this idea has given way to notions of the importance of knowledge organizations and knowledge institutions implying that favourable conditions can be arranged by mixing different competencies and organizing collaborative efforts.*

Keywords: endogenous; exogenous; knowledge society; linear model; technological frame

Kaiserfeld, Thomas. *Beyond Innovation: Technology, Institution and Change as Categories for Social Analysis.* Basingstoke: Palgrave Macmillan, 2015. DOI: 10.1057/9781137547125.0008.

This far, the ultimate test for innovations, that is, commercial success on some form of market; the ultimate focus of innovative processes on the supply side as understood, for instance, through evolutionary economics; and the ultimate attraction of innovation studies to policymakers through the claim to be able to formulate informed innovation policies, which in turn have a potentially performative value in institutions that are more efficient in supporting innovations, has been reviewed. This line of argument rests on the generally held notion that knowledge is an important engine of change in the context of institutions and technology and that engineering knowledge – stretching from earlier views on technology as applied science to more recent stress on engineering practices acquired through experience – often pointed out as crucial for technological change and as mentioned earlier in the context of definitions of technology needs to be complemented by knowledge about suitable institutions supplied by innovation studies.

In addition to the notion of innovation as satisfying market demand and innovation studies as a reliable supplier of relevant policy advice, the field of innovation studies has inherited a strive from economic growth theory, to bring technology into the theoretical thinking. Traditionally, the economic literature had stressed technological change as the consequence of factors determined by economic relations and interaction, also called endogenous factors, at the cost of exogenous factors, which are external to the theories such as access to scientific knowledge. From the 1950s, however, precursors to innovation theory developed tools to also take into account traditionally exogenous factors such as research and the learning of new practices in order to understand technological change as a factor behind economic growth. Simultaneously, the distinction between endogenous and exogenous factors was blurred by, for instance, showing how scientific knowledge, in its turn, to a large degree depended on which research problems were more easily funded compared to others.[1]

One pioneer of this perspective was the American economist Jacob Schmookler, who, in contrast to neo-classic economic theory where demand on a market creates a supply, claimed that the relations between supply and demand in effect were opposite so that supply preceded demand.[2] According to Schmookler, the primary driving force behind inventions was the use of accumulated knowledge, which he claimed to be knowledge produced due to past demand, as well as present demand for additional knowledge making scientific knowledge endogenous.³ Schmookler's notion of research as endogenous can be contrasted to Schumpeter's ideas of research for innovation as an exogenous process. Schmookler's conclusion was that "the growth of modern science and engineering is still primarily a part of the economic process".⁴ With the help of knowledge developed to satisfy demand, new innovations could be generated before demand for them existed and in this way supply preceded demand.

A model stressing exogenous scientific research has also been developed from the assumption that existing knowledge is a crucial component for loosening the restrains on invention, a notion named the linear model.⁵ The model postulated that basic research, followed by applied research and development, leads to production and diffusion and that the promotion of this chain of events is a key to supporting innovation. The general idea that empirically based and systematically produced knowledge as well as confirmed theories can be used to promote technological change as well as, for example, finding resources is old and has roots in the 16th century. However, the more specific linear model of how scientific knowledge could be exploited orderly in steps to produce inventions and later innovations was developed mainly by industrialists, consultants and economists at business schools.⁶

Behind the linear model was usually the view that scientific knowledge constituted a pool of data and competencies, which could be used, or applied, for inventions at will. The larger the pool of scientific knowledge, the better the possibilities to develop new technologies. This conclusion has more or less explicitly been the strongest argument for large-scale funding of basic research, in other words, research conducted without an aim to solve specific problems. Such reasoning has also led to notions of knowledge as one of the most important factors behind growth mirrored in concepts such as knowledge economy and knowledge society supposed to have developed during the last decades of the 20th century.⁷ Here, the commodification of knowledge has reached new heights.

Simultaneously, the linear model has been severely criticized for simplicity. For instance, supportive institutions have often been geared to the production of scientific knowledge rather than on professional skills, networking activities or other forms of backup, making it hard for entrepreneurs to exploit the knowledge they themselves judge relevant. Another strand of critique has departed from the obsession with scientific knowledge claiming that innovations more often has had their origin in practices and forms of knowledge that are far from scientific. In order to be relevant in terms of innovations, scientific knowledge, if part of the process at all, needs to be digested and edited by engineers and others practitioners to such an extent that the link between science and innovation becomes too weak to render any analytical value at all. From this perspective, a working model of relations between science and innovation cannot be linear, but stochastic at best.

Such reasoning has generated new thoughts such as those of economic historian Nathan Rosenberg. He has tried to develop concepts that more specifically point out certain features of the economic environment, such as the existing technologies or its institutions, when trying to explain technological change. In this way, he has stressed the exogenous influence of scientific and technological change while simultaneously not denying their endogenous character, especially in a world where new technologies and scientific results to a large degree depend on material resources such as laboratory equipment. Economic demand, Rosenberg argues, does not entirely decide what knowledge is acquired and what is not. There is an independent and non-negligible supply side of science and technology changing along lines determined by other factors than economic that "imposes significant constraints or presents unique opportunities which materially shape the direction and the timing of the inventive process".[8] His conclusion is that the cost of invention is different in different industries, but that it declines in general since science and technology are cumulative entities.

In more recently developed endogenous models, the claims have been somewhat weakened to statements about market incentives playing "an essential role in the process whereby new knowledge is translated into goods with practical value".[9] As a consequence, a number of models and concepts such as mode 2, post-academic science, innovation communities and triple helix have been developed with the purpose of trying to capture the dynamics between knowledge production and the use of knowledge, especially in terms of innovation and commercialization. Most of these models stress the importance of close ties, whether personal or institutional, between individuals and organizations involved in research and those involved in commercial activities.[10] A quote can illustrate the close ties between many of these models and innovation studies:

> The interaction among university, industry, and government is the key to innovation and growth in a knowledge-based economy. In ancient

Mesopotamia, a triple helix water screw, invented to raise water from one level to another, was the basis of a hydraulic system of agricultural innovation that irrigated ordinary farms as well as the Hanging Gardens of Babylon, one of the seven wonders of the ancient world. The triple helix as a physical device is succeeded by university-industry-government interactions that have led to the venture capital firms, the incubator, and the science park. These social inventions are hybrid organizations that embody elements of the triple helix in their DNA.[11]

Since the value of research is created through endogenous processes, little room seems to be left for scientific research carried out without motives of rent seeking. But it also seems as if some inventions, those that come in the form of important scientific discoveries later to be exploited, aren't as endogenous as others. There are, however, no, or at least very few, inventions that can be claimed to be altogether exogenous. To sum up, the vivid discussions on the relations between knowledge, especially scientific knowledge, and economic activities have led to insights about the problem of too far-reaching generalizations. It seems as if these relations have to be determined from case to case depending on the level of investments necessary in order to achieve relevant knowledge production as well as on the social and professional structure of the relevant industrial sector and so on.

Consider, for instance, the institution of education. It seems as if general basic training is just as important in some technological and institutional contexts as more specialized engineering education or an independent engineering profession.[12] Irrespective of the nature of engineering skills, theories regarding the knowledge of successful inventors often assume the necessity of increased specialization of knowledge and practices. Bringing these together, there are proponents of combining general basic training for the larger parts of a population with highly specialized elite groups.

From an even broader perspective, a theory of cultural development stresses the interaction within and between generations as decisive for the ability of culture to generate novelties in general and new technologies in particular. If knowledge and practices are primarily communicated across generations, for instance, by a stress on the role of grandparents in childcare through institutionalized forms of family formation, cultures tend to be conservative. In such cultures of inter-generational transmission, technological change may also be less common than in cultures of more elaborated forms of intra-generational exchange where knowledge is primarily communicated within age cohorts, for instance, by organizing education in the form of independent assignments where students have to rely more on one another than on older instructors.[13]

Later in life, such contexts tend to facilitate inventors' access to different technological frames, defined as a set of issues and knowledge in common for a relevant social group and structuring the interactions between the individuals of that group, as an important condition for his or her accomplishments.[14] More specifically, the success of inventors comes from the exclusive combination of knowledge in one specific field that proves to hold interesting clues to the solution of important problems in another field. For example, the Wright brothers used their experience as bicycle mechanics and constructors in order to solve the problems of steering airplanes.[15] The same types of ideas have been suggested to be valid on a cognitive level making the ability to invent to some extent dependent on inherited or acquired traits rather than cultural hybridity.[16] Irrespective of necessary and sufficient conditions for inventiveness, the concept of frames stresses the advantage of individuals combining knowledge from different fields, something likely to occur more frequently in a globalized world with higher levels of mobility, migration and multiculturalism – phenomena that are to a large extent dependent on both institutions, such as freedom of movement, and technologies such as access to means of transport.

In research on the sources of inventiveness, the traditional stress on individual characteristics is now more often combined with institutional influences. Ideas of technological frames can, for instance, be combined with life course perspectives analysing how "a sequence of socially defined events and roles that the individual enacts over time" can be brought together to create an understanding of how different experiences and encountered institutions influence individual behaviour on an aggregate level.[17] Correspondingly, groups of individuals with a common background can be analysed from their tendency to confront certain institutions.

In similar ways, particular situations of insight, described by their unexpectedness and sudden effortlessness, are often combined with the stressing of preparations, not the least in knowledge accumulation.[18] The view that creative uses of knowledge are more the result of an enduring process has led to many attempts to identify phases of inventive thinking.[19] Needless to say, and in line with what was stated in the second chapter when epistemic ruptures and incremental changes of technology were discussed, knowledge accumulation – both individual and collective – is more important in cases where technology and institution rely on incremental transformation of knowledge and less so in cases of its disruptive re-evaluation resulting in incommensurability, processes which have been described as paradigm shifts.[20]

Today, phases or stage models of the creative process seem dated. Instead, the focus is on multiple subprocesses of creativity such as problem finding, problem formulation, problem redefinition, generation of alternative ideas (divergent thinking), combining information, synthesis work, perception and information encoding and so on. In addition, it seems as if multiple subprocesses may be combined in different ways in order to lead to creative paths.[21] In all these subprocesses, experiences as well as acquired knowledge are important components.

As these few examples indicate, there exist cultural theories, as well as theories of individual or social problem solving, that stress the ability to either develop cognitive skills where intersecting frames or multiple subprocesses are managed, or cooperate with others from different backgrounds carrying complementary experiences, whether as a result of individual predisposition, psychological capacity or cultural practice. Over the past decades, such theories seem to have become more common and simultaneously more predictive and normative in the sense that they hold the capacity to interact over different experiences and competencies as a more or less necessary, if not sufficient, condition for creativity as well as capability to avoid or solve challenges when they occur.[22] These perspectives often stress how science and technology are concepts that merge to the limit of being indistinguishable when practices are empirically studied. It is simply not possible to distinguish scientific experimentation from technical development work: both may occur in the lab as well as in the workshop and be performed by engineers as well as scientists. In the 1950s, this insight resulted in the formation of the concept of technoscience to embrace both categories without making an analytical classification.[23] All these different ideas of how processes of technical creativity and knowledge accumulation develop are of course of interest when discussing relations between technologies, institutions and change.

Notes

1   Richard R. Nelson, "Recent Evolutionary Theorizing about Economic Change", *Journal of Economic Literature* 33:1 (March), 48–90.

2 Jacob Schmookler, *Invention and Economic Growth* (Cambridge, Mass: Harvard University Press, 1966).
3 Ibid., 175–177.
4 Ibid., 177.
5 Keith Pavitt, "Academic Research, Technical Change and Government Policy", in: *Science in the Twentieth Century*, eds, John Krige & Dominique Pestre (Amsterdam: Harwood Academic Publishers, 1997), 143–158. A critical discussion of the concept is given in: David Edgerton, "'The Linear Model' Did Not Exist: Reflections on the History and Historiography of Science and Research in Industry in the Twentieth Century", in: *The Science-Industry Nexus: History, Policy, Implications*, eds, Karl Grandin, Nina Wormbs & Sven Widmalm, Nobel Symposium 123 (Sagamore Beach, Mass: Science History Publications, 2004), 31–57; David A. Hounshell, "Industrial Research: Commentary", in: *The Science-Industry Nexus*, 59–65.
6 Benoît Godin, "The Linear Model of Innovation: The Historical Construction of an Analytical Framework", *Science, Technology, & Human Values* 31:6 (2006), 639–667.
7 Sverker Sörlin & Hebe Vessuri, "Introduction: The Democratic Deficit of Knowledge Economies", in: *Knowledge Society vs. Knowledge Economy: Knowledge, Power, and Politics*, eds, Sverker Sörlin & Hebe Vessuri (Basingstoke: Palgrave Macmillan, 2007), 1–33.
8 Nathan Rosenberg, "Science, Invention and Economic Growth", *The Economic Journal* 84:3 (March 1974), 90–108, p. 95.
9 Paul M. Romer, "Endogenous Technological Change", *The Journal of Political Economy* 98:5 (October 1990), S71–S102.
10 Michael Gibbons et al., *The New Production of Knowledge: The Dynamics of Science and Research in Contemporary Societies* (London: SAGE Publications, 1994); Henry Etzkowitz & Loet Leydesdorff, eds, *Universities and the Global Knowledge Economy: A Triple Helix of University-Industry-Government Relations* (London: Pinter Publishers, 1997); Peter Weingart, "From 'Finalization' to 'Mode 2': Old Wine in New Bottles?", *Social Science Information* 36 (1997), 591–613; Henry Etzkowitz & Loet Leydesdorff, "The Dynamics of Innovation: From National Systems and 'Mode 2' to a Triple Helix of University-Industry-Government Relations", *Research Policy* 29:2 (2000), 109–123; Helga Nowotny, Peter Scott & Michael Gibbons, *Rethinking Science: Knowledge and the Public in an Age of Uncertainty* (Cambridge: Polity Press, 2001); John Ziman, *Real Science: What It Is, and What It Means* (Cambridge: Cambridge University Press, 2001); Terry Shinn, "The Triple Helix and New Production of Knowledge: Prepackaged Thinking on Science and Technology", *Social Studies of Science* 32:4 (2002), 599–614; Helga Nowotny, Peter Scott & Michael Gibbons, "Introduction: 'Mode 2' Revisited: The New Production of Knowledge", *Minerva: A Review of Science, Learning and Policy* 41:3 (2003), 179–194.

11 Henry Etzkowitz, *The Triple Helix: University-Industry-Government Innovation in Action* (London: Routledge, 2008), 1.
12 Kees Gispen, *New Profession, Old Order: Engineers and German Society, 1815–1914* (Cambridge: Cambridge University Press, 1989).
13 Luigi Luca Cavalli-Sforza, *Genes, Peoples, and Languages* (New York: North Point Press, 1999), 179-187.
14 Wiebe E. Bijker, *Of Bicycles, Bakelites, and Bulbs: Toward a Theory of Sociotechnical Change* (Cambridge, Mass: The MIT Press, 1995), 141.
15 Tom D. Crouch, "Why Wilbur and Orville? Some Thoughts on the Wright Brothers and the Process of Invention", in: *Inventive Minds: Creativity in Technology*, eds, Robert J. Weber & David N. Perkins (Oxford: Oxford University Press, 1992), 80–92.
16 C. Scott Findlay & Charles J. Lumsden, "The Creative Mind: Toward an Evolutionary Theory of Discovery and Innovation", *Journal of Social and Biological Systems* 11:1 (1988), 3–55.
17 Janet Z. Giele & Glen H. Elder, Jr., "Life Course Research: Development of a Field", in: *Methods of Life Course Research: Qualitative and Quantitative Approaches*, eds, Janet Z. Giele & Glen H. Elder, Jr. (London: SAGE Publications, 1998), 5–27, p. 22.
18 Howard E. Gruber & S. N. Davis, "Inching Our Way Up Mount Olympus: The Evolving-Systems Approach to Creative Thinking", in: *The Nature of Creativity: Contemporary Psychological Perspectives*, ed., R. J. Sternberg (Cambridge: Cambridge University Press, 1988), 243–270; Ronald A. Finke, "Creative Insight and Preinventive Forms", in: *The Nature of Insight*, eds, R. J. Sternberg & J. E. Davidson (Cambridge, Mass: The MIT Press, 1995), 255–280.
19 Joachim Funke, "On the Psychology of Creativity", in: *Milieus of Creativity an Interdisciplinary Approach to Spatiality of Creativity*, eds, P. Meusburger, J. Funke & E. Wunder, Knowledge and Space, vol. 2 (Heidelberg: Springer Science + Business Media, 2009), 11–23.
20 Thomas S. Kuhn, *The Structure of Scientific Revolutions* (Chicago: The University of Chicago Press, 1962); Thomas Nickles, "Scientific Revolutions", in: *The Stanford Encyclopedia of Philosophy* (Summer 2014 Edition), ed., Edward N. Zalta, accessed at: http://plato.stanford.edu/archives/sum2014/entries/scientific-revolutions/, February 14, 2015.
21 Todd I. Lubart, "Models of the Creative Process: Past, Present and Future", *Creative Research Journal* 13:3–4 (2000–2001), 295–308.
22 See, for example, Alexander Styhre & Mats Sundgren, *Managing Organization Creativity: Critique and Practices* (Basingstoke: Palgrave, 2005); Thomas Heinze, Philip Shapira, Juan D. Rogers & Jacqueline M. Senker, "Organizational and Institutional Influences on Creativity in Scientific Research", *Research Policy* 38:4 (2009), 610–623.

23 Bernadette Bensaude-Vincent, *Les vertiges de la technoscience: Façonner le monde atome par atome* (Paris: La Découverte, 2009).

Further reading

Edgerton, David (2004), "'The Linear Model' Did Not Exist: Reflections on the History and Historiography of Science and Research in Industry in the Twentieth Century", in: *The Science-Industry Nexus: History, Policy, Implications*, eds, Karl Grandin, Nina Wormbs & Sven Widmalm, Nobel Symposium 123 (Sagamore Beach, Mass: Science History Publications), 31–57.

Godin, Benoît (2006), "The Linear Model of Innovation: The Historical Construction of an Analytical Framework", *Science, Technology, & Human Values* 31:6, 639–667.

7
Agency

Abstract: *Actors such as individuals or organizations are extremely important for our understanding of how knowledge and skills are converted into institutional and technological change. In some theoretical frameworks, agency is stressed as an important feature while others highlight structures of knowledge rather than individuals. Innovation studies contain both perspectives, but have traditionally cherished the will of the entrepreneur as one of the cornerstones for the understanding of change. The concepts reviewed here supply some alternative understanding of agency in the context of institution, technology and change. They range from constructivist perspectives to actor-network theory.*

Keywords: actor-network theory; biopolitics; entreprenuer; posthumanism; social constructivism; technopolitical regime

Kaiserfeld, Thomas. *Beyond Innovation: Technology, Institution and Change as Categories for Social Analysis.* Basingstoke: Palgrave Macmillan, 2015. DOI: 10.1057/9781137547125.0009.

Obviously, knowledge is an important factor in many analyses of institutional and technological change. When, moreover, trying to understand how knowledge and skills are converted into institutional and technological change or preservation, actors in the form of individuals, organizations or social groups seem an inescapable part of the equation. In some theoretical frameworks, agency is stressed as an important feature while others highlight structures of knowledge rather than individuals. Innovation studies contain both perspectives, but have traditionally cherished the will of the entrepreneur as one of the cornerstones for the understanding of change.

Already, Joseph Schumpeter underlined the individual entrepreneur's ability or, rather, an ideal type characterizing a category of actions bringing together economic, technological and scientific expertise and the use of these to get prime-mover advantages on new markets.[1] It has succinctly been described as:

> Basic inventions are more or less exogenous to the economic system; their supply is perhaps influenced by market demand in some way, but their genesis lies outside the existing market structure. Entrepreneurs seize upon these basic inventions and transform them into economic innovations. The successful innovators reap large short-term profits, which are soon bid away by imitators. The effect of the innovations is to disequilibrate and to alter the existing market structure – until the process eventually settles down in wait for the next wave of innovation. The result is a punctuated pattern of economic development that is perceived as a series of business cycles.[2]

True, Schumpeter's theory is more complex since it also includes the access to risk capital, something that in turn depends on the condition of markets. The concepts reviewed here, nevertheless, supply some alternative understanding of agency in the context of institution, technology and change.

Moreover, the early writings of Schumpeter, in which ideas of entrepreneurship and agency dominated as an exogenous force to the economic system, can be contrasted to his later writings where agency was toned down. Here, he presented innovation as more and more controlled by large organizations and enterprises, a control which in turn reinforced their ability to compete in the marketplace. These internalized and closer relations between science, technology, investment capital and the market also led to shorter time delays in innovative activities.[3]

Another, broader perspective that can be used in order to shed light on the institutional context for the development of new technologies leaving plenty of space for agency is that of social construction. There are of course a number of forms of social construction of technology.[4] The perhaps most straightforward and characteristic ones stress the importance of relevant social groups for the mobilization of science and technology as well as public participation when either supporting or opposing a technology.[5] To generalize, this perspective departs from the general notion that technology is a human product and as such something that needs to be understood from sociological and cultural perspectives rather than from material or scientific ones.

The strength of the perspective, among many different features, is that it has a potential to blur borders between producers, suppliers, consumers and users since all different categories can constitute relevant social groups although the typical situation seems to be that of producers promoting new technologies while different stakeholder groups try to mobilize resistance. Decisively, one of the most important advantages of social constructivism of technology is its debunking of any forms of predetermination of the development and use of technologies. Instead, the perspective stresses how the outcomes of design and appropriation processes are always negotiated. The room for the new reinterpretations of technologies makes it into a support for the value of activism. In many constructivist analyses, the constraints nature put on technology are thus of lesser importance. The social and institutional circumstances are considered carrying much more weight.

Simultaneously, such a model of technological change may seem simplistic, perhaps even banal. In a straightforward sense, technology is obviously socially constructed.[6] What else could it be? But with some more controversial addendums, these ideas can very easily be transformed into a much hotter substance. The most debated such addition is the assumption that scientific results as well as the world external to humanity are endogenous to social forces and relevant social groups which are able to employ these just as any other resource in their strive to promote certain designs over others. With this belief, science and technology appear as putty clay that can be modelled at will and without constraints by different actors, a notion born in the debate over science and technology as endogenous to economic factors.

Another concept mirroring institutional forces behind invention and innovation different from economic factors is technopolitical regimes or socio-technical regimes defined as coupled groups of people together with engineering and industrial practices, artefacts, political programmes and institutionalized ideologies.[7] Socio-technical regimes result from the co-evolution of institutions and technologies over time in such a way that they still contain tensions and contradictions strong enough to make the regimes dynamic.[8] Under technopolitical regimes, inventions may be driven by a strive for satisfying some culturally defined demand valued in the regime, for example, an internationally unique solution to how nuclear power can be exploited in order to produce both electricity and weapons-grade plutonium. This is called institutional logics, defined as "socially constructed, historical patterns of material practices, assumptions, values, beliefs, and rules by which individuals produce and reproduce their material subsistence, organize time and space, and provide meaning to their social reality".[9]

The main point is that concepts such as efficiency and functionality, often used when arguing for a specific technical solution in comparison to alternative ones, are extremely context-dependent. Time after time, it becomes clear how socially and culturally conditioned demands decide which inventions and innovations are created and realized although calculated and economically motivated demands may point in other directions. Functionality of an invention does not necessarily have anything to do with consumer demand or market decisions. Instead, group identity or trust, for example, may be just as important.

Both the perspectives of social construction and technopolitical regimes stress micro studies as the primary method for the understanding of dynamics between institution and technology. As historian of technology Thomas Misa has shown, to which we will return in Chapter 11, more overarching perspectives tend to lead to deterministic views on the relations between technology and institution while more detailed case studies with many different actors involved tend to lead to the stress of agency and less room for determinism. Both social constructivism and technopolitical regimes accordingly lend themselves to views of the importance of agency through a primary focus on case studies.

A third model or perspective, or perhaps method, that stresses the need to analyse agency in order to understand the relations between institution and technology, and especially their engines of change, is actor-network theory where actor and network are linked in order to avoid the problem of structure versus agency.[10] Actor-network theory

> is a concept, not a thing out there. It is a tool to help describe something, not what is being described. It has the same relationship with the topic at

hand as a perspective grid to a traditional single point perspective painting: drawn first, the lines might allow one to project a three-dimensional object onto a flat piece of linen; but they are not *what* is to be painted, only what has allowed the painter to give the impression of depth before they are erased. In the same way, a network is not what is represented in the text, but what readies the text to take the relay of actors as mediators. The consequence is that you can provide an actor-network account of topics which have in no way the shape of a network – a symphony, a piece of legislation, a rock from the moon, an engraving. Conversely, you may well write about technical networks – television, e-mails, satellites, salesforce – without at any point providing an actor-network account.[11]

In this theoretical framework, an actor is what is made to act by many others.[12] This means that even objects, animals and texts can take on agency in the form of so-called actants. Furthermore, an account, which takes these components seriously,

> is a narrative or a description or a proposition where all the actors *do something* and don't just sit there. Instead of simply transporting effects without transforming them, each of the points in the text may become a bifurcation, an event, or the origin of a new translation. As soon as actors are treated not as intermediaries but as mediators, they render the movement of the social visible to the reader. Thus, through many textual inventions, the social may become again a circulating entity that is no longer composed of the stale assemblage of what passed earlier as being part of society.[13]

Early on, underpinning the approach of actor-network theory is the claim that the perceived conventional divide between humans and artefacts as well as facts is invalid since they all have or contain agency of some immediately or distantly interacting actor. Institutions and technologies emerge and merge in networks of agency formed through the hybridization of humans, artefacts and facts where translation processes, in which mobilized resources take on their own agency, are crucial for the dynamics.[14] When a whole set of agents in the form of humans, artefacts and facts constitute a successful combination, for instance, as an established technology, fact or theory, they immediately become an invisible unit that more or less mechanically delivers output that cannot be doubted or questioned, only related to a certain input. This is the process of black boxing in which the alternatives vanish.[15] The implications are that it is impossible to keep up the division between institution and technology. Instead, they can only be understood as one and the same phenomenon consisting of actor-networks where change occurs as facts and artefacts are translated and mobilized for different purposes and ultimately to stabilize the network.

A concrete example of such a process may be the internet of things defined as a world-wide network of interconnected objects uniquely addressable, based on standard communication protocols. Although none of the parts can be called radically new, the whole network has in itself been viewed as a disruptive technology carrying many different visions of both use and abuse as it is being developed by a number of different conglomerates.[16] Autonomous cars, smart environments and abolition of theft due to tracking technologies used for any object of value are just some of the visions. One important component is radio-frequency identification making it possible to use readers to identify objects with a chip within a few metres by electromagnetic fields. The chips are small enough, about the size of a grain of rice, to be inserted into organisms such as pet animals. However, institutions to manage the far-reaching possibilities to control objects and individuals as well as collecting and supplying or hiding related data have been less developed.

The actor-network perspective is also claimed to lead to certain analytic themes. One is the stress on mobilizing resources and forging alliances, a theme which has recently been complemented by perspectives of the deconstructive aspects of actor-network theory.[17] But several others can be added. First, that cognition is situated in actor-networks in the same way as agency is. Secondly, that cognition is collective in the sense that any piece of data, claim or knowledge is distributed and relies on the efforts of several actors. Thirdly, that boundaries are fluid since networks extend beyond organizations and other types of defined entities. Fourthly, that knowledge and skills are described as material practices.[18] These themes can be used, for instance, to stress the work involved in stabilizing networks through conventions and standards. Moreover, this work may become visible when individuals, actants or cyborgs – always possessing multiple memberships of different networks – challenge practices and cognitive styles.[19] In general, actor-network analyses may be seen as an extended form of technological frames reviewed in the previous chapter on knowledge.

Donna Haraway's scholarly work is a good example of consequential network analyses with clear intentions of performativity highlighting agency among both subjects and objects, also contributing to a change of present ideals and values regarding scientific, technological and institutional practices. Her studies explore different narratives, visions as well as theories, data and information, policies and corporate practices in 20th-century technoscience.[20] Through a multi-faceted testimony, Haraway gives an associative account of how seamless webs, or rather hybrids, of institutions and practices affect things as well as animals and humans all the way down to their genes and identities. The framework is a dramatic one, which nevertheless can be influenced through action.

In common, many actor-network theorists share a stress on hybridity, not the least between body, soul and engineered devices and systems from glasses to pacemakers and stimulators of different sorts as well as chemical substances and compounds more or less sophistically researched and developed. The field is exploding with research on the relations between biotechnologies and biopolitics, from the legal and illegal global trading of organs and tissue, very much existing on market conditions, via the development of bio-objects such as embryonic stem cells from approved or unapproved lines, to the management of life through molecularization and genetization.[21] This bioengineered-techno-body has sometimes been described as a precursor of posthumanism, a condition in which agency, institutions and technologies have imploded in a realization of the cybernetic vision of the 1950s carried into the late 20th century by popular culture as well as actor-network theory and other advocates, both in business and academia.[22] These are observations we will return to in the chapters on modernism, postmodernism, hybridity and transfer.

In all these cases, just as is in the example of internet of things, methods and technologies are introduced and used without any elaborate plans for developing suitable institutions making engagement a personal matter involving high risks on an individual as well as aggregate level. For example, the trading of tissue and organs and the practice of surrogate mothers exist in a global institutional environment to some extent characterized by paternalism, exploitation and even coercion in which the legal status often enough is unclear and financial rewards can be hard, if not impossible, for donors to refuse. In this context, agency takes on a somewhat different meaning than is usually attributed to it.[23]

Yet another succinct example of underdeveloped institutions is the problem mentioned in the introduction on how to manage the spent nuclear fuel generated by the operation of nuclear power plants. There are two distinct solutions to the problem. On one hand, spent nuclear fuel can be loaded with agency by keeping the door open for use and reuse in Gen IV reactors. On the other, spent nuclear fuel can be bereaved of agency by deep geological disposal using technical barriers and other means to secure its seclusion from the world of agency. In its most extreme forms, not even monitoring of the radioactive waste buried deep underground is being planned since even the passive observation of its transformation may potentially load it with agency to incorporate it in the networks of future generations.

Notes

1. Jon Elster, *Explaining Technical Change: A Case Study in the Philosophy of Science* (Cambridge: Cambridge University Press, 1983), 126.
2. Richard N. Langlois, "Schumpeter and the Obsolescence of the Entrepreneur", in: *Austrian Economics and Entrepreneurial Studies*, eds, Roger Koppl, Jack Birner & Peter Kurrild-Klitgaard, Advances in Austrian Economics 6 (Bingley: Emerald Publishing, 2003), 283–298, p. 285.
3. Christopher Freeman, *The Economics of the Industrial Innovation*, 2nd ed. (London: Pinter Publishers, 1982). For a critique of the division between an early and a late Schumpeter, see Langlois, "Schumpeter".
4. Sergio Sismondo, "Some Social Constructions", *Social Studies of Science* 23:3 (1993), 515–553; Peter Taylor, "Co-construction and Process: A Response to Sismondo's Classification of Constructivisms", *Social Studies of Science* 25:2, 348–359.
5. Wiebe E. Bijker, Thomas P. Hughes & Trevor Pinch, eds, *The Social Construction of Technological Systems* (Cambridge, Mass: The MIT Press, 1989); Wiebe E. Bijker & John Law, eds, *Shaping Technology/Building Society: Studies in Sociotechnical Change* (Cambridge, Mass: The MIT Press, 1992); Sergio Sismondo, *Science without Myth: On Constructions, Reality, and Social Knowledge* (Albany: State University of New York Press, 1996); Ian Hacking, *The Social Construction of What?* (Cambridge, Mass: Harvard University Press, 1999).
6. Langdon Winner, "Upon Opening the Black Box and Finding It Empty: Social Constructivism and the Philosophy of Technology", *Science, Technology, & Human Values* 18:3 (1993), 362–378.
7. Gabrielle Hecht, *The Radiance of France: Nuclear Power and National Identity after World War II* (Cambridge, Mass: The MIT Press, 1998). The concept of technopolitical regimes relies on technical regimes and their leading to concrete technopolitical phenomena as discussed in: Langdon Winner, "Techne and Politeia: The Technical Constitution of Society", in: *Philosophy and Technology*, eds, Paul T. Durbin & Friedrich Rapp (Dordrecht: Kluwer, 1983), 97–111.
8. Lea Fuenfschilling & Bernhard Truffer, "The Structuration of Socio-technical Regimes – Conceptual Foundations from Institutional Theory", *Research Policy* 43:4 (2014), 772–791.
9. Ibid., 775.

10 John Law, "After ANT: Complexity, Naming and Topology", in: *Actor Network Theory and After*, eds, John Law & John Hassard (Oxford: Blackwell, 1999), 1–14.
11 Bruno Latour, *Reassembling the Social: An Introduction to Actor-Network Theory* (Oxford: Oxford University Press, 2005), 131.
12 Ibid., 46.
13 Ibid., 128.
14 Bruno Latour & Steve Woolgar, *Laboratory Life: The Construction of Scientific Facts* (London: SAGE Publications).
15 Katinka Waelbers & Philipp Dorstewitz, "Ethics in Actor Networks, or: What Latour Could Learn from Darwin and Dewey", *Science and Engineering Ethics* 20:1 (2013), 23–40.
16 Luigi Atzori, Antonio Iera & Giacomo Morabito, "The Internet of Things: A Survey", *Computer Networks* 54:15 (2010), 2787–2805.
17 Vasilis Galis & Francis Lee, "A Sociology of Treason: The Construction of Weakness", *Science, Technology, & Human Values* 39:1 (2014), 154–179.
18 Susan Leigh Star, "The Trojan Door: Organizations, Work, and the 'Open Black Box'", *Systems Practice* 5:4 (1992), 395–410.
19 Susan Leigh Star, "Power, Technology and the Phenomenology of Conventions: On Being Allergic to Onions", in: *A Sociology of Monsters: Essays on Power, Technology and Domination*, ed., John Law, Sociological Review Monographs 38 (London: Routledge, 1991), 26–56.
20 Donna J. Haraway, *Modest_Witness@Second_Millennium. FemaleMan©_Meets_ OncoMouse™: Feminism and Technoscience* (London: Routledge, 1997).
21 Catherine Waldby & Robert Mitchell, *Tissue Economies: Blood, Organs and Cell Lines in Late Capitalism* (Durham, NC: Duke University Press, 2006); Niki Vermeulen, Sakari Tamminen & Andrew Webster, eds, *Bio-Objects: Life in the 21st Century* (London: Ashgate, 2012); Nikolas Rose, *Politics of Life Itself: Biomedicine, Power and Subjectivity in the Twenty-First Century* (Princeton: Princeton University Press, 2006).
22 Langdon Winner, "Resistance Is Futile: The Posthuman Condition and Its Advocates", in: *Is Human Nature Obsolete? Genetics, Bioengineering, and the Future of the Human Condition*, eds, Harold W. Baillie & Timothy K. Casey (Cambridge, Mass: The MIT Press, 2005), 385–410.
23 Catherine Waldby & Robert Mitchell, *Tissue Economies: Blood, Organs and Cell Lines in Late Capitalism* (Durham, NC: Duke University Press, 2006).

Further reading

Haraway, Donna J. (1997), *Modest_Witness@Second_Millennium. FemaleMan©_Meets_OncoMouse™: Feminism and Technoscience* (London: Routledge).

Latour, Bruno (2005), *Reassembling the Social: An Introduction to Actor-Network Theory* (Oxford: Oxford University Press).

Leigh Star, Susan (1991), "Power, Technology and the Phenomenology of Conventions: On Being Allergic to Onions", in: *A Sociology of Monsters: Essays on Power, Technology and Domination*, ed., John Law, Sociological Review Monographs 38 (London: Routledge), 26–56.

Winner, Langdon (1983), "Techne and Politeia: The Technical Constitution of Society", in: *Philosophy and Technology*, eds, Paul T. Durbin & Friedrich Rapp (Dordrecht: Kluwer), 97–111.

8
Clusters, Systems and Blocks

Abstract: *An early attempt to analyse processes emanating from combinations and integration of knowledge together with its carriers stemmed from the observation that industrial branches and sectors seemed to agglomerate geographically, creating local or regional clusters. Other concepts mirror sectorial complexes or the importance of systems. In common, they all have the assumption that technology and institutions are systemic in the sense that their parts cannot be understood or analysed in isolation, but need to be understood as connected entities. Conclusions in this chapter include the developing uniformity of systems and organizations despite the original institutional differences of distinctive geographical locations and regions. Another insight is that change in one part of the complex is very likely to have repercussions throughout the whole system.*

Keywords: bottleneck; development block; multiple invention; regional cluster; reverse salient; simultaneous invention; socio-technical system

Kaiserfeld, Thomas. *Beyond Innovation: Technology, Institution and Change as Categories for Social Analysis.* Basingstoke: Palgrave Macmillan, 2015.
DOI: 10.1057/9781137547125.0010.

From the emphasis of innovation studies and many other frameworks for analyses of institutions, technologies and change on the importance of combinations of different forms of knowledge and skills as well as the cooperation of actors in order to convert that knowledge into change (or preservation) follows the recommendation in the name of performativity to make institutions facilitate the communication and cooperation between vocations, professions and other groups as well as individuals representing different experiences and competencies to be involved in technological change. A number of concepts have been developed in order to explain processes emanating from combinations and integration of knowledge as well as its carriers. The example of actor-network theory has already been described at some length.

One important perspective in the vein of understanding complementing abilities has been that of economic geography where Alfred Marshall early on pointed out that industrial branches and sectors seemed to agglomerate geographically, creating local or regional clusters including relevant services and subcontractors, labour markets and institutions supporting appropriate competencies.[1] Marshall's observation of agglomeration has been combined with new growth theory of the 1980s and 1990s where it is stated that certain types of investments, specifically those made in human capital, raise output and growth that can be sustained also over extended time periods due to prime-mover advantages and temporary monopoly effects.[2] Taken together, they indicate that regional knowledge production may be a way to create a higher and sustained productivity in industry and trade of geographically defined areas, at least if some coordination is achieved between firms and different forms of infrastructure resulting in a cluster.[3]

From this perspective, universities and other establishments of higher education, especially engineering schools, can be expected to boost regional innovation activities making it to some extent discouraging when a Dutch economist found that regions surrounding Dutch universities failed to show lasting effects on other indicators of economic growth than income and job opportunities.[4] The general ideas of agglomeration and new growth theory were, however, strengthened by another study of different regions in the United States. Here it was concluded that a strong correlation existed between university research and innovations in larger metropolitan regions with a population exceeding one million and with more than 30,000 students enrolled in higher education.[5]

Geographical aspects including population densities have also been combined with other social characteristics in order to explain why some regions seem to demonstrate more intense innovation activities than other. The perhaps best-known and at least most criticized attempt has been the notion of the creative class, roughly constituted by skilled labour and professionals with a pinch of bohemians, which can thrive when gathering in predominantly urban environments where tolerance is combined with high-tech cultures.[6] Although this specific attempt has been extremely questioned, for instance, due to method and data analyses, the general idea of combining geographical and social as well as professional features in order to characterize successful clusters of innovation is a popular innovation studies theme.

A somewhat different take on the problem of why some regions seem to generate more new technologies than others departs from the idea that technology itself transforms through combinations of existing technological building blocks into new technologies that are again combined. In this way, new more complex technologies are formed on existing components of simpler ones leading to more complex and more diverse forms. A consequence is that an innovation made in one area may be applied to another set of problems. In regions where different competences exist, the preconditions are better to develop the potentials of technologies across different areas of expertise.[7]

A more institutional approach still stressing the interaction between different organizations and actors more than rules and regulations and traditional practices is the notion of systems of innovation, a concept that has been described as one of the most important developed within innovation studies over the past decades.[8] Innovation systems come in different shapes and are most commonly defined by geographical scope or industrial branch. Thus, they are usually national, regional or sectorial. But no matter of attribute, this is in essence an institutional perspective on innovations stressing both the interdependence between different actors involved in innovation activities such as firms, individuals, public authorities, special interest groups etc. and the fact that innovations and perhaps also inventions, more often than not, emerge in the intersection between different organizations and fields of knowledge where public authorities also are thoroughly engaged.[9] In order to understand why some inventions become innovations, maybe later also to be diffused, it is necessary to map and analyse the relations between different organizations as well as the legal, social and economic institutions that guide their exchanges and actions such as cooperation or competition. It should perhaps be added that these ideas most effectively describe invention and innovation of capital goods where close cooperation between producer and consumers can be essential, or innovations appearing in technology procurement projects where new, not yet existing, technologies are ordered on behalf of some public buyer, typically a national defence organization.

The concept of development blocks is also used to illustrate the importance of links between different actors, activities and institutions as a condition for technological change with the difference that it is not as closely linked to geographical vicinity as clusters and systems usually are. A development block is the factors linked to a specific industrial activity. The growth of a development block depends on the complementary investments made in other fields related to it. As a result, imbalances and structural tensions may appear within the development block as in socio-technical regimes mentioned in the previous chapter, which in turn may cause further changes and invention activities. Imbalances can arise for different reasons, either by market signals; through a drive for efficiency; or by changes in network relations between firms and other organizations. They may be the result of activities within a single firm or of cooperation between a number of actors.[10] The key point is that imbalances cause both depression and expansion depending on their type.

But not only different institutional aspects facilitate the combination of knowledge and practices. Also technology has important systemic aspects. It is, for instance, nonsensical to buy a fax machine if no other potential senders and receivers have one. But if some decide to buy a fax and the use diffuses, their user value increases and soon enough there may even develop a pressure to acquire a fax among some of those earlier reluctant to buy one. These phenomena are called network effects or network externalities.[11] One common network externality is increased production followed by lowered prices per item due.

Other technologies may, conversely, be more beneficial to the user the less compatible they are. When secrecy is demanded, for instance, in military contexts, it is a virtue to keep the number of machines for generating and deciphering code to a minimum although there need to be at least two and in reality even more in order for the system to function. A case in point is the electro-mechanical rotor cipher machines used during World War II to cipher and decipher code.[12] Otherwise, original machines benefitting from singularity are mostly different types of production machines that rely on exclusiveness in order to keep the prices of the products up. An example of this type of machines is the difference engines developed from the 1780s in order to calculate, tabulate and print results of polynomial functions. The ideas on how to design and construct them were further developed in England during the 1820s, and a rather efficient type of difference engine with a printing device attached to it was the Scheutzian calculation engine of the 1850s, a construction further developed in the 1870s.[13]

Most commonly, however, technologies relying on systems have peak efficiency when there are enough compatible versions to make it beneficial to develop and maintain an institutional and technological system for their use while ever growing numbers may lead to saturation and negative effects on the system. Automobile systems are a good example of this type of development. Automobile use has advantages as long as the number of car users is limited. Extensively distributed use, however, leads to mass automobility with detrimental consequences despite efforts to build roads with higher capacity or to introduce institutions to limit traffic such as pay roads or congestion charges. In the long run, it seems hard to maintain efficiency in a system of boundlessly growing automobility without limiting accessibility or accepting inefficient personal transport due to traffic congestion. Thus, there may be an optimal network size and eventually, excessive use may make the system less efficient despite efforts to expand it.

In all these different examples, there are close interactions between technological and institutional conditions that rely on the systemic nature of technology, that is, that technological artefacts and devices are not isolated, but rely on supportive networks and systems. In more general terms, this insight has resulted in the concept of sociotechnical systems including a model for development through different overlapping phases.[14] The first of these contains invention and development of the system, much through the efforts of inventors. The second stage is characterized by technological transfer from one region or society to others. The third phase is system growth and expansion when efficiency is improved. At this stage, it is necessary that system development is coordinated so that the weakest links, whether technical or institutional, do not disturb general performance. In the fourth phase, competition between systems and within systems develops. And in the fifth phase, finally, there is consolidation in order to conserve resources.

More interesting from the perspective of institutional and technological change is perhaps the consequences of the systematic character of different parts of socio-technical systems. The connectedness of technology implies that it will work only as well as its weakest link. With this insight in mind, weak links or bottlenecks, also denoted reverse salients borrowing a military term for resistance pockets in an advancing front, in combination with market forces may create enormous demand for new technological solutions to come to terms with specific bottlenecks that prevent the socio-technical system from being more efficient.[15] In other words, if a link of a system seems to lag behind other parts functionally, there will be very high (demand-driven) incentives to improve or replace it with something that again can match the output of other parts of the system.

Therefore, according to socio-technical-system theory, inventors, managers, investors and other agents work hard to identify bottlenecks in the system, which predominantly occur during the third phase of system growth and expansion. These weaknesses are then transformed into critical problems to be solved by experts, engineers, scientists, legal experts or financial analysts. The higher the potential for growth in the system and the more important the identified bottleneck is for its efficiency, the higher the demand for a solution of the resulting critical problem. Even more important is the point that a critical problem doesn't need to be technical, but can be institutional demanding different types of experts to be solved in order to avoid system inefficiency. No matter how the idea is labelled, the common denominator is the notion of a crucial problem that, if solved, with high certainty will generate system growth and profit or public benefit. The result is often a massive funnelling of resources for the solution of the critical problem resulting in among other things the phenomenon of multiple or simultaneous invention.

One example of this dynamic is the development of incandescent light bulbs with metal filament from the 1880s when electrical systems were starting to be introduced in different cities in order to compete with existing systems for gas light.[16] Efficient electrical light sources were soon identified as a reverse salient and a lot of efforts were put in to develop an efficient and cheap light bulb with a longer lifetime than existing electrical lamps. In order to solve this critical problem, large electrical companies such as Westinghouse and Siemens hired physicists and chemists to establish industrial research labs. A vast number of new patents for different light bulbs were also filed in the early decades of the

20th century eventually resulting in light bulbs with tungsten filament used for light bulbs up until today.[17] The pressing demand for a solution to the critical problem had concrete scientific and technical solutions in the form of a metal filament in which a main challenge was to engineer production techniques for the filament, which relied on a ductile form of tungsten, and institutional solutions manifested in new laboratories at electrical companies.

Another important concept used to understand the dynamics of socio-technical systems is technological momentum. Momentum can be understood as a measure of the efforts needed to transform the system, or parts of it, once established and usually become salient after the system has grown and expanded during the third phase. The reason is that during the earlier stages of system development, its survival hinges on relatively few enthusiasts, whether users or inventors, making its technological momentum relatively low. As the system grows and its critical problems are solved, institutions are arranged, for instance in terms of formalized education and legal settlements as well as standards, the system gains momentum. The more institutions are arranged and engaged, the harder and more costly it will become to transform the system.[18]

Descriptions of technological momentum are paralleled by observations in business history where the 19th-century development of technologies, for instance, relying on steam and electricity, has been claimed to result in business-management innovations such as accounting and financing as well as the creation of middle management and a corresponding social "managerial class".[19] Such new elaborate forms of management were a prerequisite for mass production and mass innovation on national markets and eventually global markets and have been used to explain the surpassing in economic performance by the United States compared to the United Kingdom in the late 19th and early 20th centuries.[20] Here, organizational changes and their resulting consequences are seen as the key explanation for the creation of institutional momentum. The stress on a transformed business organization complements a concept like technological momentum in a way necessary to fully appreciate the importance of analysing institutional and technological change in tandem.[21]

An important conclusion is the developing uniformity of systems and organizations despite the original institutional differences of distinctive geographical locations and regions. According to theory of socio-technical systems, this homogenization of socio-technical systems can be explained by market demands on efficiency. Bringing the different ideas reviewed in this chapter together, two additional conclusions can be drawn. First, that a complex of technologies and institutions can't be more efficient than its weakest link or part. Second, that changes in one part of the complex is very likely to have repercussions throughout the whole system. This last insight is perhaps one of the most important when explaining the difficulties in developing an accepted theory of technology, institution and change as well as the existence of so many different attempts and perspectives.

Notes

1 Alfred Marshall, *Industry and Trade* (London: Macmillan, 1919).
2 Paul M. Romer, "Endogenous Technological Change", *The Journal of Political Economy* 98:5 (October), S71–S102.
3 Maryann P. Feldman, *The Geography of Innovation*, Economics of Science, Technology and Innovation 2 (Dordrecht: Kluwer, 1994); Mark Muro & Bruce Katz, "The New 'Cluster Moment': How Regional Innovation Clusters Can Foster the Next Economy", in: *Entrepreneurship and Global Competitiveness in Regional Economies: Determinants and Policy Implications*, eds, Gary D. Libecap & Sherry Hoskinson, Advances in the Study of Entrepreneurship, Innovation & Economic Growth, 22 (Bingley: Emerald Publishing, 2011), 93–140.
4 Raymond Florax, *The University: A Regional Booster* (Aldershot: Avebury, 1992).
5 Attila Varga, *University Research and Regional Innovation: A Spatial Econometric Analysis of Academic Technology Transfer* (Dordrecht: Kluwer, 1998).
6 Richard Florida, *The Rise of the Creative Class* (New York: Basic Books, 2002); Richard Florida, *Cities and the Creative Class* (New York: Routledge, 2005).
7 W. Brian Arthur, *The Nature of Technology: What It Is and How It Evolves* (New York: The Free Press, 2009).
8 Ben R. Martin, "Innovation Studies: An Emerging Agenda", in: *Innovation Studies: Evolution and Future Challenges*, eds, Jan Fagerberg, Ben R. Martin & Esben S. Andersen (Oxford: Oxford University Press, 2013), 168–186.
9 Bengt-Åke Lundvall, ed., *National Systems of Innovation: Towards a Theory of Innovation and Interactive Learning* (London: Anthem Press, 1992); Charles Edquist, ed., *Systems of Innovation: Technologies, Institutions and Organizations* (London: Pinter Publishers, 1997); David Doloreux, "What We Should Know about Regional Systems of Innovation", *Technology in Society* 24:3 (2002),

242–263; Frank W. Geels, "From Sectoral Systems of Innovation to Sociotechnical Systems: Insights about Dynamics and Change from Sociology and Institutional Theory", *Research Policy* 33:6–7 (2004), 897–920.
10. Håkan Lindgren, ed., *Economic Dynamism*, The Institute for Economic and Business History Research, Research Report No. 6 (Stockholm: Stockholm School of Economics).
11. S. J. Liebowitz & Stephen E. Margolis, "Network Externality: An Uncommon Tragedy", *Journal of Economic Perspectives* 8:2 (1994), 133–150.
12. David Kahn, *Seizing the Enigma: The Race to Break the German U-Boats Codes, 1939–1943* (Boston: Houghton Mifflin, 1991).
13. Michael Lindgren, *Glory and Failure: The Difference Engines of Johann Müller, Charles Babbage and Georg and Edvard Scheutz*, Linköping Studies in Arts and Science 9, Stockholm Papers in History and Philosophy of Technology 2017 (Linköping: Linköping University, 1987).
14. Thomas P. Hughes, *Networks of Power: Electrification of Western Society, 1880–1930* (Baltimore: Johns Hopkins University Press).
15. Thomas P. Hughes, "The Dynamics of Technological Change: Salients, Critical Problems, and Industrial Revolutions", in: *Technology and Enterprise in a Historical Perspective*, eds, Giovanni Dosi, Renato Gianetti & Pier Angelo Toninelli (Oxford: Oxford University Press, 1992), 97–118.
16. Hans-Joachim Braun, "Gas oder Elektrizität? Zur Konkurrenz zweier Beleuchtungssysteme, 1880–1914", *Technikgeschichte* 47 (1980), 1–19.
17. Arthur A. Bright, Jr., *The Electric-Lamp Industry: Technological Change and Economic Development from 1800 to 1947* (New York: Macmillan, 1949); Harold C. Passer, "Development of Large-Scale Organization: Electrical Manufacturing Around 1900", *The Journal of Economic History* 12:4 (1952), 378–395; George Wise, "A New Role for Professional Scientists in Industry: Industrial Research at General Electric, 1900–1916", *Technology and Culture* 21:3 (1980), 408–429; George Wise, "Ionists in Industry: Physical Chemistry at General Electric, 1900–1915", *Isis* 74:1 (1983), 7–21; George Wise, *Willis R. Whitney, General Electric, and the Origins of U.S. Industrial Research* (New York: Columbia University Press, 1985).
18. Thomas P. Hughes, "Technological Momentum", in: *Does Technology Drive History? The Dilemma of Technological Determinism*, eds, Merrit Roe Smith & Leo Marx (Cambridge, Mass: The MIT Press, 1994), 101–114.
19. Alfred J. Chandler, *The Visible Hand: The Managerial Revolution in American Business* (Cambridge, Mass: Harvard University Press, 1977).
20. Alfred Chandler, *Scale and Scope: The Dynamics of Industrial Capitalism* (Cambridge, Mass: Harvard University Press, 1990).
21. David A. Hounshell, "Hughesian History of Technology and Chandlerian Business History: Parallels, Departures and Critics", *History and Technology* 12:3 (1995), 205–224.

Further reading

Doloreux, David (2002), "What We Should Know about Regional Systems of Innovation", *Technology in Society* 24:3, 242–263.

Geels, Frank W. (2004), "From Sectoral Systems of Innovation to Socio-technical Systems: Insights about Dynamics and Change from Sociology and Institutional Theory", *Research Policy* 33:6–7, 897–920.

Hughes, Thomas P. (1994), "Technological Momentum", in: *Does Technology Drive History? The Dilemma of Technological Determinism*, eds, Merrit Roe Smith & Leo Marx (Cambridge, Mass: The MIT Press), 101–114.

Muro, Mark & Bruce Katz (2011), "The New 'Cluster Moment': How Regional Innovation Clusters Can Foster the Next Economy", in: *Entrepreneurship and Global Competitiveness in Regional Economies: Determinants and Policy Implications*, eds, Gary D. Libecap & Sherry Hoskinson, Advances in the Study of Entrepreneurship, Innovation & Economic Growth, 22 (Bingley: Emerald), 93–140.

9
Resistance to Change

Abstract: *An important perspective on institutions and technology is how they resist change, isolated or in tandem. Categories of institutional change are reviewed together with concepts such as path dependency, technological momentum and increasingly costly reversibility, all capture processes in which material and institutional practices and norms are stable. Resistance to change seems to rely on two fundamental characteristics. In some cases, it can be derived from the costs involved when changing practices or concretely substituting old technologies for new. In others, individual behaviour may conserve existing practices, for instance, through conscious reluctance to change or through the force of routine. Most often, however, a combination of the two is the most important prerequisite for resistance to change.*

Keywords: historical institutionalism; increasingly costly reversibility; isomorphism; new institutionalism; path dependence; technological momentum

Kaiserfeld, Thomas. *Beyond Innovation: Technology, Institution and Change as Categories for Social Analysis*. Basingstoke: Palgrave Macmillan, 2015. DOI: 10.1057/9781137547125.0011.

By introducing the concept of socio-technical systems and technological momentum, this survey has slowly started to review ideas of the relations between institution, technology and change that are not part of the core literature of innovation studies, but instead constitute alternatives. To generalize, one important such complement is the different notions of how institutions and technology, isolated or in tandem, resist change. Admittedly, such ideas are also addressed in innovation studies, most commonly perhaps through the concept of path dependence, which will be discussed in detail in the following paragraphs. In this chapter, however, technological and institutional elements of resistance will be reviewed and compared more systematically. These elements will furthermore be highlighted, developed and generalized beyond concepts such as socio-technical regimes and development blocks used to understand resistance to change as they occur in specific institutional structures.

On a fundamental level, the Danish economist Ester Boserup has stressed the demographic environment when analysing new agricultural technologies. She has claimed that new methods and technologies for growing provisions are developed only under pressure of lacking resources which occur when the population grows to such an extent that existing methods and technologies do not suffice to supply the food needed using the land at hand. New and more efficient methods and technologies are developed in order to balance food supply with the rising need on a limited area of land. The novelties occur at the cost of intensified farming in terms of higher productivity per land unit, which also implies lower productivity per labour hour in traditional agricultural settings. Therefore, knowledge of how to increase production is a necessary though not sufficient condition for the introduction of new farming technologies. In addition to knowledge, demographic pressure leading to lack of food is also necessary for the introduction of new inventions.[1] Boserup is unique in the sense that she does not view the possibility to invent as a sufficient condition for presenting new technologies. In addition, she claims the need of an imperative force in order to have changes in agricultural methods, namely, demographic pressure. This is very different from the ideas that incentives for innovation created, for instance, in socio-technical systems are so overwhelming that no extra imperative force is needed in order to explain its occurrence.

Another form of resistance relevant here is that type described by Rosenberg when pointing out how existing technologies in order to meet competition from new radical innovations often go through stages of incremental improvements.[2] Rosenberg's example is the development of the iron hull cargo steam ship in the mid-19th century, which caused improvements of sailing ships in the 1860s and 1870s. In a sense, this observed sudden innovative activity to meet new radical innovation with incremental improvements of existing technologies could be seen as a form of resistance to innovation too.

Although seldom applied in the context of technology, theoretical approaches of new institutionalism are interesting when trying to shed light on continuity and stability rather than change in general institutional settings. The strand of political science called new institutionalism is sometimes described through a number of slogans. Here, it is simply seen as a research tradition with an interest in how formal and informal institutions shape human action. In comparison to traditional institutionalism, new institutionalism does not adhere to ideas that institutions can be understood as instruments for human intervention or that they must be related to contemporary social structures. Instead, the basic notion is that institutions carry weight and influence the outcomes of human agency by their own force, not only in the sense that they constitute constraints or generate opportunities, but also in the formation of individual identities and preferences.[3] Institutions are key intervening variables and to some extent perhaps even independent.

Many of the concepts reviewed here, though not all, can be connected to a form of new institutionalism called historical institutionalism, a tradition that emphasizes the power of institutional tradition and explains change, to generalize broadly, by pointing to external shock or the build-up of institutional tensions. Historical institutionalism has also paid attention to continuous change. Within this amorphous tradition, ideas and belief systems have recurrently been pointed out as crucial for the formation of policies, best practices and other forms of institutions.[4] From the perspective of historical institutionalism, the effects of institutions may be described as unintended and even against rational choice.[5]

Traditionally, new institutionalism plays down the role of individuals and when individuals nevertheless act, they do it in contexts of existing opportunities and change. Instead, new institutionalism stresses the stability of institutions and suggests persisting, constant and incremental changes, rather than windows of opportunities during unsettled periods. Thus, there are notable similarities with theories of evolutionary economics although new institutionalism perhaps puts a greater stress on history. A consequence is that the idea of path dependence becomes an important notion, a concept we will return to later on in this chapter. The incremental change usually stressed in new institutionalism can still be categorized. Wolfgang Streeck and Kathleen Thelen have identified patterns in terms of displacement, layering, drift, conversion and exhaustion.[6] Displacement is a type of change where an institution is moved in an institutional system to assume new assignments and its actors new behaviour. The institution is either transformed within or through invasion by actors from the outside. Obviously, new technology can displace institutions, for instance, when databases become more accessible and commonly used on account of printed books and journals making libraries more like central subscription service units than deposits of printed material.

Another form of institutional change is layering, a process where new institutions are developed but with the same purpose and function as older, existing ones. For instance, electronic news sites making news footage quicker and more easily accessible have recently been layered on printed newspapers, which still exist but also have to adapt and find new ways of selling their services. A process of layering may also be connected to the notion of sedimentation, which has been developed to denote the imagined last stage of a process of institutionalization. Sedimentation occurs after an exogenous institutional innovation has been habituated, made permanent and spread. It "fundamentally rests on the historical continuity of structure, and especially on its survival across generations of organizational members".[7] In other words, it can be characterized by "perpetuation of structures over a lengthy period of time".

Thirdly, institutions may drift which means that they slowly assume new structures and meanings through neglect of institutional maintenance, for instance, when engineering schools transform teaching practices to be more relevant in the scientific laboratory than on the workshop floor.[8] This is an example of academic drift, conventionally defined as a process entailing an increased valuation and assimilation of academic practices.[9] This can be contrasted to the reverse development, epistemic drift, defined as processes by which values from ideological systems external to science, for instance, business or policy, are adopted by researchers making them pay increasing attention to the potential uses of their activities and practices.[10] When organizations drift while new ones are formed to replace them, a specific form of layering occurs.[11]

Conversion is the fourth form of institutional change when an existing institution assumes new goals, functions or purposes. A case where technology may have led to conversion is the institution of international standards. Earlier, an important purpose of this institution was to safeguard the different standard units and to use them to distribute national copies to different states. But as these material objects have shown to vary in mass, length and so on, the standard units have been redefined in terms of fundamental constants of nature. Today, only the International Prototype Kilogram in the outskirts of Paris remains the definition of an international standard. The institution of international standards has slowly conversed to see as its main purpose carrying out measurement-related research and defining new standards in terms of natural constants.

The fifth form of institutional change has been named exhaustion and relates more to breakdown and death than to change. One example where new technology has led to exhaustion is the advent of automobile society making the institutions supporting horse and carriage traffic in the cities to have been exhausted. The first four of these categories have been further elaborated on and given characterizations departing from whether new rules are introduced (in displacement and layering) or if the impact or enactment of older rules have been changed while they still are, at least formally, valid (drift and conversion).[12]

Another very different way of understanding institutions is to view them as systems of rules, beliefs, norms and organizations that taken together generate regular behaviour among agents and in this way accomplish an equilibrium. Stable institutions are thus nothing less than equilibrium patterns of behaviour. From this perspective, institutional change occurs when expectations are changed with the advantage that formal and informal rules are equally important. It is the way they affect behaviour and thus institutional stability that decide their analytical value. Obviously, an exogenous factor such as technology may disrupt equilibrium like the use of guns threatened samurai warfare in 17th-century Japan leading to regulations against its unlimited use.[13] Such occasions are sometimes called critical junctions.

An already mentioned concept, in the third chapter on market institutions, that is common when analysing institutional change is isomorphism. Originally, organizational isomorphism was seen as a result of evolutionary selection of efficiency on organizational traits in organizations existing under similar environmental conditions. In an important article,

Paul J. DiMaggio and Walter Powell claimed that isomorphism could evolve for other reasons as well, such as legitimacy. Interacting organizations form an organizational field in which it may be legitimating to converge in terms of practice and form, a process leading to institutional isomorphism.[14] Organizations can be coerced into isomorphism by other organizations they depend on or through expectations. Organizations may also mimic other organizations in situations of ambiguity or confusion in order not to take unnecessary risks at times of uncertainty. Thirdly, institutional isomorphism may be generated by normative pressure as when norms are developed among agents in an organizational field through similar education or professional networks.

DiMaggio and Powell then went on to formulate a number of hypotheses regarding different circumstances that may strengthen isomorphic tendencies, both in organizations and organizational fields. One of these hypotheses claims that demand for centralized resources may drive isomorphism in the sense that the dependent organization will transform to resemble the organization that controls the resources in demand. In the cases where the resources demanded are technical, this influences organizational convergence and generates isomorphism. One example to illustrate this process of isomorphism is how 21st-century information and communication technologies have made virtually all public authorities and agencies in welfare economies expected to provide 24-hours services including access to electronic forms and information as well as registration of applications and so on.[15]

One concept widely used over the past decades, and often connected to new institutionalism, is path dependence.[16] This is also the concept that has been most widely used in innovation studies when characterizing resistance to change. Change is path dependent if there is a sequence of events where "important influences upon the eventual outcome can be exerted by temporally remote events, including happenings dominated by chance elements rather than systematic forces".[17] Simultaneously, no one would claim that it is possible to decide the direction or path in the growth of technological knowledge merely by reference to certain initial conditions. Rather, one needs to consider the history of the system.[18] The concept was originally defined to describe dynamics in economic changes and exemplified by the keyboard layout of early typewriters where the topmost row of letters read QWERTY from left to right. The argument was that the use of the same layout in modern computer keyboards should be understood as a result of path dependence rather than "systematic forces". Soon, however, critics showed that closer readings of the empirical material exposed a case far from decisive and that there were sound reasons to argue that the QWERTY layout was indeed efficient also in comparison to other layouts.[19]

The debate was nevertheless important and path dependence was also related to other similar concepts such as excess inertia appearing when the switch to new standards was resisted due to the supposed benefit from compatibility.[20] According to one economic model, where actors are identical and decide sequentially whether to change to a new technology they all benefit from, it is rational for each to change if they all have complete information. If, however, information flow is incomplete, there will always be excess inertia and the situation becomes more complicated. In one case – symmetric inertia – all actors would benefit from the new technology if they all change, but they do still not make the change since they wait for others to take initiative since each and everyone does not benefit individually if no other follows their example. In this case, status quo is maintained and the equilibrium will be disturbed only if one or a few actors change technology despite initial costs and risks. The suggested solution is coordination between a number of actors making the change beneficial to them and others who may follow. In the other case – asymmetric excess – the actors differ in their evaluation of benefits from a change of technology although total benefits exceed total costs. In this case, it is possible that only some of the actors change technology with the result of dispersion of standards.

At the heart of the debate lay the question if it is possible that a minor advantage for a "technology, product, or standard can have important and irreversible influences on the ultimate market allocation of resources, even in a world characterized by voluntary decisions and individually maximizing behaviour".[21] Using the example of video standards, VHS and Betamax, economists have proved that only under important restrictions on prices, institutions and foresight are inefficiencies due to initial conditions not remedied. The claim that dependence on initial conditions leads to an inefficient outcome despite being remediable is a stronger version of path dependence, which is very unlikely to occur in neo-classical economic models. Crucial in these models is thus again the information different actors have on different occasions, and assuming that actors have necessary information about possible future developments and the costs for reversing them, they will avoid undesirable lock-in effects. And when they nevertheless occur, the stability of inefficient equilibria relies on high costs for reversibility.

The concept of increasingly costly reversibility departs from the notion that there are two opposite ideal conditions concerning the costs for reversing a process. On the one hand, one could imagine a situation of costless reversibility in which processes without any additional costs can be reversed at a specific point. On the other hand, one can imagine the opposite situation of complete irreversibility where a process may not be reversed irrespective of the resources spent.[22] It is important to remember that these are ideal conditions and that in reality, there are always costs of some sort involved when reversing a process, if nothing else at least the loss of time. Simultaneously, it is hard to imagine a situation, which is completely irreversible, in the sense that it cannot be reversed at any given cost, at least not if reversibility is assumed to be weak, that is a process characteristic that allows for the process to be brought back to a given state without having to follow the same trajectory of intermediary states as is the case for strong reversibility.[23]

The concepts reviewed so far have all been characterized from the perspective of technological and institutional change or stability. When used to shed light on the relations between institutions, technology and change, they generally stress how technologies function either as a catalyst that facilitates institutional change, for example, through some radical invention leading to the formation of socio-technical systems (later on stabilized through technological momentum) or a new market, or as an obstacle to institutional change by strengthening path dependence or making reversibility more costly. Path dependence, technological momentum and costly reversibility are all, in one way or another, used to point out the large efforts often needed to liberate a conglomerate of technology and institution from its history.

Moreover, when trying to formulate a generalizable feature of the relations between technology and institution from a materialistic point of view, it seems tempting to regard the most important consequences of technology's materiality as the constraints it puts on institutional change. Bridges are occasionally built. Sometimes, they happen to collapse. But not counting military actions, it is extremely rare that they are purposely torn down without being replaced by a new bridge. It is namely through the characteristics of materiality in terms of demand for large investments to make it function in a specific institutional setting that technology contributes to path dependence, gains momentum and accelerates the increasingly costly reversibility. From a materialistic point of view, it may thus seem as if technology constrains institutions more than the other way around.

On the contrary, institutions can be claimed to be just as path dependent, exposed to momentum and costly reversibly also when they have nothing to do with technology at all. At least, this is the established wisdom of new institutionalism where the perspective is that institutions are almost as historically stabilizing as if they had themselves been cast in concrete. From this perspective, assuming that technology moulds institutions into especially stabilized forms would be to jump to conclusions. Instead, the concepts reviewed should be understood as highlighting phenomena that can hinder or impede both technological and institutional change.

To summarize, the ideas reviewed here seem to rely on two fundamental characteristics. In some cases, resistance can be derived from the costs involved when changing practices or concretely substituting old technologies for new. In others, individual behaviour may conserve existing practices, for instance through conscious reluctance to change or through the force of routines, and thus collectively generate what can be perceived as resistance. In most cases, however, a combination of the two is the most likely prerequisite for resistance, for instance realized in discouragement due to the time and resources needed in order to transform existing institutions or technologies.

Notes

1   Ester Boserup, *The Conditions of Agricultural Growth: The Economics of Agrarian Change under Population Pressure* (London: Allen & Unwin, 1965).
2   Nathan Rosenberg, *Perspectives on Technology* (Cambridge: Cambridge University Press, 1976), 202-206.
3   André Lecours, "Introduction", in: *New Institutionalism: Theory and Practice*, ed., André Lecours (Toronto: University of Toronto Press, 2005), 3-25.
4   B. Guy Peters, *Institutional Theory of Political Science: The "New Institutionalism"*, 2nd ed. (London: Continuum, 2005), 71-86.
5   Ellen M. Immergut, "The Theoretical Core of the New Institutionalism", *Politics & Society* 26:1 (1998), 5-34; Peter A. Hall, "Historical Institutionalism in Rationalist and Sociological Perspective", in: *Explaining Institutional Change: Ambiguity, Agency, and Power*, eds, James Mahoney & Kathleen Thelen (Cambridge: Cambridge University Press, 2010), 204-223.

6   Wolfgang Streeck & Kathleen Thelen, "Introduction": Institutional Change in Advanced Political Economies", in: *Beyond Continuity: Institutional Change in Advanced Political Economies*, eds, Wolfgang Streeck & Kathleen Thelen (Oxford: Oxford University Press, 2005), 3–39.
7   Pamela S. Tolbert & Lynne G. Zucker, "The Institutionalization of Institutional Theory", in: *Handbook of Organization Studies*, eds, S. Clegg, C. Hardy & W. Nord (London: SAGE Publications, 1996), 175–190, p. 184.
8   Jonathan Harwood, "Understanding Academic Drift: On the Institutional Dynamics of Higher Technical and Professional Education", *Minerva: A Review of Science, Learning and Policy* 48:4 (2010), 413–427.
9   For definition and thorough review, see ibid.
10 Aant Elzinga, "Research, Bureaucracy and the Drift of Epistemic Criteria", in: *The University Research System: The Public Policies of the Home of Scientists*, eds, Björn Wittrock & Aant Elzinga, Studies in Higher Education in Sweden No 5. (Stockholm: Almqvist & Wiksell, 1984), 191–220; Aant Elzinga, "The Science-Society Contract in Historical Transformation with Special Reference to 'Epistemic Drift'", *Social Science Information* 36:3 (1997), 411–445.
11 Thomas Kaiserfeld, "Why New Hybrid Organizations Are Formed: Historical Perspectives on Epistemic and Academic Drift", *Minerva: A Review of Science, Learning and Policy* 51:2 (2013), 171–194.
12 James Mahoney & Kathleen Thelen, "A Theory of Gradual Institutional Change", in: *Explaining Institutional Change: Ambiguity, Agency, and Power*, eds, James Mahoney & Kathleen Thelen (Cambridge: Cambridge University Press, 2010), 1–37.
13 Noel Perrin, *Giving Up the Gun: Japan's Reversion to the Sword, 1543–1879* (Boston: Godin, 1979). This monograph has been criticized for neither relying on enough resources, nor attributing developments to acknowledged circumstances, see: Conrad Totman, "Review", *Journal of Asian Studies* 39:3 (1980), 599–601.
14 Paul J. DiMaggio & Walter W. Powell, "The Iron Cage Revisited: Institutional Isomorphism and Collective Rationality in Organizational Fields", *American Sociological Review* 48:2 (1983), 147–160.
15 Lemuria Carter et al., "E-Government Utilization: Understanding the Impact of Reputation and Risk", *International Journal of Electronic Government Research* 8:1 (2012), 83–97.
16 Paul A. David, "Clio and the Economics of QWERTY", *The American Economic Review* 75:2 (May 1985), 332–337.
17 Ibid., 332.
18 Nathan Rosenberg, *Exploring the Black Box: Technology, Economics, and History* (Cambridge: Cambridge University Press, 1994), 10; James Mahoney, "Path Dependence in Historical Sociology", *Theory and Society* 29:4 (2000), 507–548.

19 Jean-Philippe Vergne, "QWERTY Is Dead; Long Live Path Dependency", *Research Policy* 42:6–7 (2013), 1191–1194.
20 Joseph Farrell & Garth Saloner, "Standardization, Compatibility, and Innovation", *The RAND Journal of Economics* 16:1 (1985), 70–83.
21 S. J. Liebowitz & Stephen E. Margolis, "Path Dependence, Lock-In, and History", *The Journal of Law, Economics, & Organization* 11:1 (1995), 205–226.
22 Andrew B. Abel & Janice C. Eberly, "Optimal Investment with Costly Reversibility", *The Review of Economic Studies* 63:4 (1996), 581–593.
23 Patrick Suppes, "Weak and Strong Reversibility of Causal Processes", *Stochastic Causality*, eds, M. C. Galavotti, P. Suppes & D. Constantini (Stanford: CSLI Publications, 2001), 203–220.

Further reading

DiMaggio, Paul J. & Walter W. Powell (1983), "The Iron Cage Revisited: Institutional Isomorphism and Collective Rationality in Organizational Fields", *American Sociological Review* 48:2, 147–160.

Lecours, André, ed. (2005), *New Institutionalism: Theory and Practice* (Toronto: University of Toronto Press).

Liebowitz, S. J. & Stephen E. Margolis (1995), "Path Dependence, Lock-In, and History", *The Journal of Law, Economics, & Organization* 11:1, 205–226.

Mahoney, James (2000), "Path Dependence in Historical Sociology", *Theory and Society* 29:4, 507–548.

Mahoney, James & Kathleen Thelen, eds. (2010), *Explaining Institutional Change: Ambiguity, Agency, and Power* (Cambridge: Cambridge University Press).

Vergne, Jean-Philippe (2013), "QWERTY Is Dead; Long Live Path Dependency", *Research Policy* 42, 1191–1194.

10
Commons

Abstract: *A specific theoretical framework in which a technological component has added value and increases precision of the analyses is that of common property resource management, commons for short. Two general problems have been identified with common-pool resources, overuse and free riders that may benefit from the resource without having to share the costs for its use. In general, commons lead to complex governance structures where the understanding of change in both commons and technologies is improved when the two entities are brought together.*

Keywords: commons; envirotechnical regime

Kaiserfeld, Thomas. *Beyond Innovation: Technology, Institution and Change as Categories for Social Analysis*. Basingstoke: Palgrave Macmillan, 2015.
DOI: 10.1057/9781137547125.0012.

New institutional theory and other important notions of institutional characteristics, such as equilibrium and isomorphism, are relevant for discussions on institutional and technological change although the analyses they generate seldom take technologies into account as an endogenous factor. In order to do this, other concepts need to be invoked that will describe specific versions of interaction between institutions and technologies. Here follows a discussion on a specific theoretical framework in which a technological component has added value and increases precision of the analyses. If resistance to change could be claimed as a part of innovation studies, not the least by the application of the concept of path dependence, the theoretical framework of commons that will be discussed here is yet another step further away from the core literature of innovation studies.

By adding technology to the analyses of commons, or common property resource management as the term has been named formally, this set of influential theories about institutional stability and change has been developed considerably over the past years. A common property resource or common-pool resource is a resource from which it is hard to exclude users, for example, the air we breathe, and it has been defined as "a valued natural or human-made resource or facility that is available to more than one person and subject to degradation as a result of overuse. Common-pool resources are ones for which exclusion from the resource is costly and one person's use subtracts from what is available to others."[1] This last restriction is important since it excludes knowledge and its different materializations such as databanks from being common-pool resource. Two general problems have been identified with common-pool resources, overuse and free riders that may benefit from the resource without having to share the costs for its use.

A simple example would be a river where upstream dams for the extraction of hydropower for watermills or water-powered sawmills makes it harder to use the resource further down the stream. In order to distribute water and power more evenly, the dam constructions as well as regulation of water flow need to be managed through some sort of institutional frameworks. Of course, technologies to some extent determine the salient features of the river, and as technologies change, appropriate institutions handling the management of the river as a resource are likely to change too. With the help of concepts such as envirotechnical regimes and landscapes, where the interface between nature and technology interacts with ideologies and institutions to underpin envirotechnical systems, this reasoning can be generalized.[2]

Central to the notion of commons is thus the idea that nature supplies resources such as schools of fish, herds of cattle, arable land, flowing water, blowing wind, wood and ore, to name just a few, that humans can exploit, harvest and share. From a historical perspective, exploitation of natural resources seems to have escalated in Europe during the process of industrialization. To some extent, this may be attributed to a higher rate of population growth. Simultaneously, a dominating religious view of our relations to the environment has also been highlighted as an explanation. The hypothesis is that the management of our natural resources depends on our views on the relation between them and ourselves involving notions of a common destiny for our natural environment and us. The Judaeo-Christian tradition narrates a linear worldview with a clear beginning and a clear end. In it, the world has furthermore been created for the benefit of humanity who in turn is assigned to study nature and name its content in order to expose the creation of God. Taken together, this religiously based ideology – system of thoughts, ideas and truth-beliefs – made it legitimate to keep developing technologies and institutions that could be used to further exploit natural resources with increased efficiency without taking into account the threat of depletion.[3]

The basic problems then include the co-development and tuning of technologies and institutions in order to make exploitation possible, for instance, by rewarding investments of time and effort, while simultaneously securing future access, for instance, by limiting it. This notion has been further developed in a model of multilevel perspective where social-ecological system is the unit of analysis typically rooted in a particular spatial context. It moves through different states that are dependent on multilevel interactions with niche, regime and landscape.[4]

In the original analyses of the problems connected to commons and the institutions developed to handle common property resource management, technology seldom played any important part. Instead, the variables often used to predict the outcome of collective cooperation in the context of a common-pool resource include participants and their positions, possible actions, information, control, potential outcomes and analyses of net costs and benefits.[5] The involvement of technology in this equation was more implicit.

More recently, however, there has been much interest on not only how technologies are used in the management of commons, but also how they can span commons and indeed decide and define what becomes a common-pool resource. Historian of technology, Nina Wormbs, has shown how bandwidth in the electromagnetic spectrum became a common-pool resource when public national radio broadcasting was developed in Europe in the mid-1920s.[6] Intriguing are the technical complexities involved when trying to distribute frequencies among different European nations and decide the number of radio transmitters to be allowed in each of them. In addition, this common-pool resource is transformed as the power and precision of transmitters are developed.

Another example of a common-pool resource is international airspace, again not only dependent on, but in fact constituted by technologies. The motivation behind the reduction of national security, implied by allowing foreign airplanes to fly across a nation's own skies, was a desire to develop faster travelling by the use of air flight across international borders.[7] In the process of states negotiating to open their sovereign airspace to commercial or public airlines operated in the interests of other countries, a multitude of issues had to be solved from luggage-allowance rules to weather services. In the case of airspace, where the common-pool resource is unchanged by the user, technologies have nevertheless been developed in order to allow shorter space and time between planes sequentially using portions of the common. Obviously, institutions are powerful enough not only to manage commons in an efficient way, but also to supply incentives to develop management technologies increasing its quantity.

The tendency that a number of common-pool resources evolved at a time of both formation of states and state power as well as the development of strong technological networks has had consequences. In general, commons lead to "complex governance structures, with governments, intergovernmental organizations, private organizations, and public non-governmental organizations mutually shaping and participating in governance regimes, often with no direct involvement by individual citizens".[8] The insight that the understanding of change in both commons and technologies is improved when the two entities are brought together is important when we now turn to concepts designed to more specifically deal with technological and institutional change.

Notes

1   Thomas Dietz, Nives Dolšak, Elinor Ostrom & Paul C. Stern, "The Drama of the Commons", in: *The Drama of the Commons: Committee on the Human*

Dimensions of Global Change, eds, Elinor Ostrom et al. (Washington, D.C: The National Academic Press, 2002), 3–35, p. 18.
2. Sara B. Pritchard, *Confluence: The Nature of Technology and the Remaking of the Rhône* (Cambridge, Mass: Harvard University Press, 2011). See also: William Cronon, *Nature's Metropolis: Chicago and the Great West* (New York: W.W. Norton, 1991).
3. Lynn White, Jr., "The Historical Roots of Our Ecological Crisis", *Science* 155:3767 (March 1967), 1203–1207.
4. Adrian Smith & Andy Stirling, "The Politics of Social-Ecological Resilience and Sustainable Socio-technical Transitions", *Ecology and Society* 15:1 (2010), accessed at: http://www.ecologyandsociety.org/vol15/iss1/art11/, May 20, 2014.
5. Ellinor Ostrom, *Understanding Institutional Diversity* (Princeton: Princeton University Press, 2005).
6. Nina Wormbs, "Technology-Dependent Commons: The Example of Frequency Spectrum for Broadcasting in Europe in the 1920s", *International Journal of the Commons* 5:1 (2011), 92–109.
7. Eda Kranakis, "The 'Good Miracle': Building a European Airspace Commons, 1919–1939", in: *Cosmopolitan Commons: Sharing Resources and Risks across Borders*, eds, Nil Disco & Eda Kranakis (Cambridge, Mass: The MIT Press, 2013), 13–55.
8. Nil Disco & Eda Kranakis, "Toward a Theory of Cosmopolitan Commons", in: *Cosmopolitan Commons: Sharing Resources and Risks across Borders*, eds, Nil Disco & Eda Kranakis (Cambridge, Mass: The MIT Press, 2013), 57–96.

Further reading

Disco, Nil & Eda Kranakis, eds (2013), *Cosmopolitan Commons: Sharing Resources and Risks across Borders* (Cambridge, Mass: The MIT Press).
Ostrom, Elinor et al., eds (2002), *The Drama of the Commons: Committee on the Human Dimensions of Global Change* (Washington, D.C: The National Academic Press).

11
Technological Determinism

Abstract: *Innovation studies usually shun the idea that technology influences institutional settings in a more or less predetermined way, a notion mirrored in the concept technological determinism, which in turn exists in many different versions. One important set of deterministic theories is of course different forms and shades of Marxism. Others are materialistic theories with notions of predetermined developments although not necessarily in a revolutionary form. Today, determinism is seldom encountered in scholarly literature as a main road to exciting analyses of relations between technological and institutional change. Exceptions may, however, be found in visions of technological developments following certain quantitative regularities such as Moore's law. But these are more speculative predictions than theoretical frameworks for improved understanding of technological and institutional change.*

Keywords: eotechnics; Marxism; Moore's law; multiple invention; neotechnics; paleotechnics; simultaneous invention; soft determinism; strong determinism; technological determinism; transformational invention; weak determinism

Kaiserfeld, Thomas. *Beyond Innovation: Technology, Institution and Change as Categories for Social Analysis*. Basingstoke: Palgrave Macmillan, 2015. DOI: 10.1057/9781137547125.0013.

The idea that technology influences institutional settings in a more or less predetermined way is often called technological determinism, a concept which comes in many different versions.[1] In its strongest form, technological determinism is seen as a stance where technology determines institutional contexts and is itself predetermined, for example, through a strive for increased efficiency or productivity. Such notions have been underpinned by observations of simultaneous invention suggesting determining relations between established institutions and new technologies such that a specific institutional context, for instance, an accumulation of knowledge regarding a particular area, implies a definable set of inventions.[2] From this perspective, a technological novelty or transformation is latent in, or determined by, the existing institutional and technological ensemble. Note, however, that this is an understanding of multiple or simultaneous invention, which is broader than the explanation given in the eighth chapter where it was understood as a consequence of massive funnelling of resources for the solution of specific problems of high relevance for system development.

Somewhat weaker is the notion that technology determines institutional structures, but not necessarily it being predetermined itself, which leaves room for the possibility that social institutions have influenced technical change in earlier stages. One prominent example of this type of determinism is the infamous claim of Lynn White, Jr., that the stirrup was the key to the evolution of feudal society in Europe from the 8th century.[3] This situation resembles institutional lag, which was referred to in the second chapter on concepts of technology and institution. Even weaker is the interpretation that there are unintended consequences of technological change such as the breaking up of traditions and social order when introducing new technologies among indigenous peoples such as steel axes among Aboriginal Australians or snowmobiles among Finnish Skolt Sàmi.[4] Another scale used to establish stronger and weaker forms of technological determinism departs from how important technology is assumed to be in dynamic processes. In cases where technology is the key factor behind change, determinism is considered to be stronger. When technology is just one component in a set of factors, technological determinism is weaker.[5]

In addition, there is the less rigid idea often referred to as soft determinism where existing technologies are thought to span a space of opportunities of which some are picked depending on different institutional or cultural factors.[6] But the notion of opportunity space may also be made more complex by observing it from the individual's perspective. From this angle, it may often seem to shrink rather than expand in high-tech societies due to spillover effects of new technologies:

> Business economists [–] have failed to observe that as the carpet of "increased choice" is being unrolled before us by the foot, it is simultaneously being rolled up behind us by the yard. We are compelled willy-nilly to move into the future that commerce and technology fashion for us without appeal and without redress. In all that contributes in trivial ways to his ultimate satisfaction, the things at which modern business excels, new models of cars and transistors, prepared foodstuffs and plastic *objets d'art*, electric tooth-brushes and an increasing range of push-button gadgets, man has ample choice. In all that destroys his enjoyment of life, he has none. The environment about him can grow ugly, his ears assailed with impunity, and smoke and fuel gases exhaled over his person. He may be in circumstances that he will never enjoy a night's rest at home without planes shrieking overhead. Whether he is indifferent to such an invasion of his privacy, whether he suffers it stoically or painfully, whether he is resigned or furious, there is under the present dispensation practically nothing he can do about it.[7]

Note that the limits of alternatives (or the accessible opportunity space) discussed by Edward Mishan does not depend on exogenous technology, but can instead be interpreted as accounted for by (implicit) institutional factors surrounding technological externalities. Another version of this notion is the description of technology and institution as being a tool for some actors to influence or even prescribe the behaviour, feelings, tastes and even ethical or political outlook of another larger group of users. This is sometimes formulated as a script inscribed in a technology, or perhaps more commonly a product, by designers for users to follow.[8]

When discussing institution, technology and change in terms of determinism, the perspective of time is needless to say of key importance. As historian of technology Thomas Misa has shown, technological determinism is more likely to appear in some analytical perspectives than in others.[9] There is thus a tendency in analyses of technology, institution and change that extend over longer time periods and take into account more overarching perspectives such as general social tensions or economic growth, to highlight technology as an exogenous factor that drives institutional change. Correspondingly, there is a tendency in analyses that consider dynamics in greater detail and over shorter time periods, perhaps involving a more limited number of actors in an institutional setting where a radical invention surfaces, to stress institutional context as an exogenous factor that influences technological design. Temporality and level of analysis is apparently an extremely important factor for the choice of perspective when studying technological and institutional change.

Accordingly, there are a number of important theories dealing with the dynamics of institution, technology and change over very long time periods resulting in more or less deterministic conclusions, for instance, Marxism and many of its derivatives. The difference in comparison to evolutionary economics is that Marxism is revolutionary in the sense that the important changes are much more dramatic and occur during time periods of revolution and upheaval. Marxism is definitely a theory of technological and institutional change ultimately derived from the experiences of the industrial revolution in Europe in general and Britain in particular. It is nevertheless important to point out that changes analysed within Marxist theory traditionally develop over longer time periods than only a few decades as the typical changes analysed within evolutionary economics.

Karl Marx argued that capitalists innovate because they are forced to do so by competition, and they are able to innovate because they can draw on an accumulated stock of inventions, that is, on science.[10] So far, his ideas in principle resemble those of evolutionary economic theory. On the contrary, Marx introduces the concept of modes of production consisting of two distinct categories: forces of production, often interpreted as physical technologies including available tools, instruments and machines together with all sources of energy, and relations of production often interpreted as social institutions. Marx argues that forces of production over time depart from being in correspondence with the relations of production. This process implies that forces of production sooner or later will be in contradiction to the relations of production. These contradictions may take many forms such as crisis or lead to too many restrictions on changes in the forces of production.

The general problem for those interested in Marx's theory of invention is that the factors behind the changes in forces of production are described differently in different texts by Marx. In some, it is claimed that contradictions between forces and relations of production appear only when all productive forces for which there is room (within a set of relations) have been developed.[11] In other texts, innovative activities are regarded as springing from individual inner sources of being. If so, the problem is not creating incentives for innovation, but removing obstacles.¹² Generally, earlier texts of Marx stress utopian visions derived from subjective and psychological states while later ones are viewed as more technical and scientific in their description of the relations between forces of production and relations of production. To sum up, Marxian theories come in many different shapes and colours nevertheless making Marxism a set of theories, which in general take both technological and institutional change seriously, although sometimes with a strong tendency towards determinism.¹³ This tendency is reflected in the following very famous quote from the younger Marx.¹⁴

> In acquiring new productive forces men change their mode of production; and in changing their mode of production, in changing the way of earning their living, they change all their social relations. The hand-mill gives you society with the feudal lord; the steam-mill, society with the industrial capitalist.¹⁵

Although Marx can obviously be quoted on deterministic statements, there are just as many other instances where he has given very different perspectives on the historical relations between technologies and institutions. There are thus many versions of materialistic, albeit less deterministic, theories of institutional and technological change in the vein of Marxism, a theme I will return to later on.

A cultural historian who has also stressed the existing technological level as the most important environment when analysing the development of new technologies is Lewis Mumford. More specifically, he has discussed how dominating materials have limited the possible alternatives for technicians of different times. Mumford's hypothesis was that machine culture, or civilization as he termed it, has developed in phases and that builders and artisans of the different phases had turned to specific and different areas for inspiration and building material.¹⁶ In the eotechnic era (1000–1800), primarily wood and organic fibres were used to build machines, vehicles and constructions such as houses and bridges. Feather pens were used for writing. Materials were organic, thus the term eotechnic coming from earth. In the eotechnic era, energy accordingly came from muscles, water and wind. Romanticizing the pastoral past, Mumford also claimed that cultures strove for a harmonious balance between the senses and the freedom from labour.

Technically civilized lifestyles were set off with the clock, the most important basis for the development of capitalism since it made time fungible (and thus transferable). The clock became the prototype for all machines. The second phase, the paleotechnic (roughly 1700–1900), was "an upthrust into barbarism, aided by the very forces and interests which originally had been directed toward the conquest of the environment and the perfection of human nature".[17] Inventions of the paleotechnic were made trying to solve specific problems rather than hunting for general scientific principles; in fact, scientific learning was devalued by businessmen. The invention of coal-fired steam-powered factories and the installation of capital-intensive machinery led to a gigantic round-the-clock scale of production supported by unskilled machine tenders. Mumford identified iron as the primary building material of the paleotechnic, and skyscrapers, bridges and steamships as primary accomplishments. War and mass sport were social releases from mechanized life, and the hysteric duties of wartime production were a natural outgrowth of the tensions and structures of paleotechnic life.

In describing the neotechnic age (from about 1900 to Mumford's present, 1930), Mumford focused on the invention of electricity. He furthermore saw the neotechnic phase as dominated by scientists rather than mechanically apt machinists. Scientists were concerned with the invisible, the rare, the atomic level of change and innovation. Compact and lightweight aluminium was the metal of the neotechnic and the management of communication and information primary accomplishments.

In his reasoning about technical phases, Mumford attributed internal characteristics to specific materials influencing methods of engineering and artisan work. In the eotechnic phase, agriculture and handicrafts such as sloyd dominated and the family was the most common production entity, in its turn a consequence of eotechnics. In the paleotechnic phase, clocks and time was the most important machinery made out of metals and giving this phase its character of mass production in factories. Neotechnics was characterized by the mixing of materials and methods leading to the control of different phenomena such as electricity liberating production geographically and later on also time-wise.

Important technological shifts relied on the transit between the three phases. In order to explain change, however, Mumford was less original and for the eotechnical era, he insisted that existing technologies and knowledge production were important generators for new technologies, while consumption of luxuries among the wealthy laid the foundations for markets and demand when they spread to other layers of society. Hardly a very original thought in the shadow of

Karl Marx and Werner Sombart. In conclusion, Mumford seems to have relaxed his determinism when explaining shifts in phases while defending it when describing dynamics in each of the three different phases he observed.

Today, determinism is seldom encountered in scholarly literature as a main road to exciting analyses of relations between technological and institutional change and especially not in innovation studies. There are to be sure exceptions such as visions of technological developments following certain quantitative regularities such as Moore's law regarding the number of transistors on integrated circuits projected to double approximately every second year, but these are to be considered speculative predictions more than theoretical frameworks for improved understanding of technological and institutional change.[18] Simultaneously, present theories of institutional change in the social sciences and economics less often involve technology as an endogenous factor than they did 30–40 years ago and this despite the fact that many report an impression of technologies reshaping everyday life on a global scale more than they did then. Marxism has had its heydays and while technology is more often recognized as a potential world changer, institutional analyses, as we have seen, still tend to treat it is an exogenous force with evolutionary economics as the notable exception. There often seems to be room for determinism in one form or another, although this is seldom acknowledged or even understood.

The most recent indication of this is perhaps the debated hypothesis that the future holds no transformational inventions or innovations that can be compared to the industrial revolutions occurring over the past 250 years due to the development of different generic technologies: steam power, electrical power and internal combustion and, most recently, the distribution of information and communication technologies.[19] The result is predicted to be a much lower growth making the past 250 years a parenthesis of extreme economic development and distribution of welfare. Against this deterministic hypothesis, it is of course common to point out market institutions as a historically efficient method for finding profitable technologies with strong candidates being medical technologies or technologies developed to save the different limited resources of today's earth.[20] Interestingly enough, the debate seems to always focus on new technologies rather than new institutions, for instance, developed to distribute growth differently in the light of a lower frequency of transformational technologies.

Notes

1. Sally Wyatt, "Technological Determinism Is Dead: Long Live Technological Determinism", in: *The Handbook of Science and Technology Studies: Third Edition*, eds, Edward J. Hackett et al. (Cambridge, Mass: The MIT Press, 2008), 165–180.
2. A. L. Kroeber, *Anthropology: Culture Patterns and Processes* (New York: Harcourt, Brace & World, 1963), 149–175.
3. Lynn White, Jr., *Medieval Technology and Social Change* (Oxford: Clarendon Press, 1962). For an example of a critical review, see: P. H. Sawyer & R. H. Hilton, "Technical Determinism: The Stirrup and the Plough", *Past & Present* 24:1 (April 1963), 90–100.
4. Paul S. Adler, "Technological Determinism", in: *International Encyclopedia of Organization Studies*, eds, Stewart R. Clegg & James R. Bailey (London: SAGE Publications, 2008), 1537–1540; Lauriston Sharp, "Steel Axes for Stone-Age Australians", *Human Organization* 11:2 (1952), 17–22; Pertti J. Pelto, *The Snowmobile Revolution: Technology and Social Change in the Arctic* (Menlo Park, Calif: Cummings Publishing, 1973).
5. Brian Martin, "Technological Determinism Revisited", *Metascience* 4:2 (1995), 158–160.
6. Nathan Rosenberg, *Exploring the Black Box: Technology, Economics, and History* (Cambridge: Cambridge University Press, 1974), 15; Leo Marx & Merrit Roe Smith, "Introduction", in: *Does Technology Drive History? The Dilemma of Technological Determinism*, eds, Merrit Roe Smith & Leo Marx (Cambridge, Mass: The MIT Press, 1994), ix–xv.
7. Edward J. Mishan, *Growth: The Price We Pay* (London: Staples Press, 1969), 52.
8. Madeleine Akrich, "The De-Scription of Technical Objects", in: *Shaping Technology/Building Society: Studies in Sociotechnical Change*, eds, Wiebe E. Bijker & John Law (Cambridge, Mass: The MIT Press, 1992), 205–224.
9. Thomas J. Misa, "Retrieving Sociotechnical Change from Technological Determinism", in: *Does Technology Drive History? The Dilemma of Technological Determinism*, eds, Merrit Roe Smith & Leo Marx (Cambridge, Mass: The MIT Press, 1994), 115–141.
10. Quoted from: Elster, *Explaining Technical Change: A Case Study in the Philosophy of Science* (Cambridge: Cambridge University Press, 1983), 166.
11. Ibid., 211–212.
12. Ibid., 216.
13. For a view of Marx as a neo-classical thinker, see: Nathan Rosenberg, "Marx as a Student of Technology", in: *Inside the Black Box: Technology and Economics* (Cambridge: Cambridge University Press, 1982), 34–51.
14. Bruce Bimber, "Three Faces of Technological Determinism", in: *Does Technology Drive History? The Dilemma of Technological Determinism*, eds,

Merrit Roe Smith & Leo Marx (Cambridge, Mass: The MIT Press, 1994), 79–100.
15   Karl Marx, *The Poverty of Philosophy*, original in French in 1847 (London: Martin Lawrence, 1937), 92.
16   Lewis Mumford, *Technics and Civilization* (New York: Harcourt, Brace & Company, 1934).
17   Ibid., 154.
18   For an example, see: Raymond Kurzweil, *The Age of Spiritual Machines* (New York: Viking Press, 1999); Michio Kaku, *The Physics of the Future: How Science Will Shape Human Destiny and Our Daily Lives by the Year 2100* (New York: Doubleday, 2011). On Moore's law, see: Robert R. Schaller, "Moore's Law: Past, Present, and Future", *IEEE Spectrum* 34:6 (June 1997), 52–59; Chris A. Mack, "Fifty Years of Moore's Law", *IEEE Transactions on Semiconductor Manufacturing* 24:2 (May 2011), 202–207.
19   Robert J. Gordon, "Is U.S. Economic Growth Over? Faltering Innovation Confronts the Six Headwinds", NBER Working Papers 18315 (Cambridge, Mass: National Bureau of Economic Research, August 2012), accessed at: http://faculty-web.at.northwestern.edu/economics/gordon/Is%20US%20 Economic%20Growth%20Over.pdf, June 30, 2014.
20   Thomas B. Edsall, "No More Industrial Revolutions?", *The New York Times*, October 15, 2012, accessed at: http://campaignstops.blogs.nytimes. com/2012/10/15/no-more-industrial-revolutions/?_php=true&_type=blogs& module=Search&mabReward=relbias%3Ar%2C%5B%22RI%3A5%22%2C%22 RI%3A18%22%5D&_r=0, June 30, 2012.

Further reading

Smith, Merrit Roe & Leo Marx, eds (1994), *Does Technology Drive History? The Dilemma of Technological Determinism* (Cambridge, Mass: The MIT Press).

Wyatt, Sally (2008), "Technological Determinism Is Dead: Long Live Technological Determinism", in: *The Handbook of Science and Technology Studies: Third Edition*, eds, Edward J. Hackett et al. (Cambridge, Mass: The MIT Press), 165–180.

12
Modernity and Its Critics

Abstract: *The nebulous concept of modernism has often been connected to notions of determinism in the sense that modernity follows a perceived trajectory of technological progress towards greater measures of artificiality and control, urbanity and rationality. In this complex, transformed notions of speed and space as well as possibilities and threats have often been highlighted. Modernity is thus a double-edged sword. Strive for control of nature, as well as the understanding of humanity's role as part of nature, has been the centre of the problem. Although modernity has underscored the emancipative force of what seems to have been institutionally distributed growth due to technical change, modern thinkers have to an increasing extent stressed how complexes of institutions and technologies limit individual and institutional freedom.*

Keywords: autonomy; cognitive artefact; modernity; risk society; secularization; technocracy; technological imperative; urbanization

Kaiserfeld, Thomas. *Beyond Innovation: Technology, Institution and Change as Categories for Social Analysis*. Basingstoke: Palgrave Macmillan, 2015. DOI: 10.1057/9781137547125.0014.

Although determinism is less popular today than it has been, widely applicable theories of technological and institutional change where agency has usually been toned down are still valid approaches. One such framework is known under the heading of modernity. This nebulous concept has often been connected to determinism through the idea that institutional change follows a perceived trajectory of technological progress towards greater measures of artificiality and control, urbanity and rationality.[1] Another salient feature of modernity inherited from Enlightenment has been the conviction that any question has only one correct answer and that the world could be controlled and ordered "if we could only picture and represent it rightly".[2] In this complex, transformed notions of speed and space as well as possibilities and threats have often been highlighted.[3] Modernity is thus a double-edged sword where early promoters hoped for emancipation through control of nature and development of both material and moral circumstances while 20th-century intellectuals often have stressed the catastrophic consequences of what has been understood as humanity's lacking control of rationality and its instruments.[4]

These themes have been recurring in the literature on technological and institutional change. For instance, in the 1960s economist John Kenneth Galbraith claimed that "we are becoming the servants in thought, as in action, of the machine we have created to serve us".[5] Or in the words of political scientist Langdon Winner

> Although virtually limitless in their power, our technologies are tools without handles. Often they seem to resist guidance by preconceived goals or standards. Far from being merely neutral, our technologies provide a positive content to the area of life in which they are applied, enhancing certain ends, denying or even destroying others.[6]

A strive for control of nature, as well as an understanding of humanity's role as part of nature after secularization terminated the Judaeo-Christian tradition of a linear worldview that had up until then legitimated exploitation of natural resources, has been at the centre of the problem. The assumption of many modern thinkers has accordingly been the existence of a novel single correct mode of technological representation supplying the instruments needed to fulfil dreams of control or an autonomous force threatening human existence in general or, possibly, both at the same time.

Langdon Winner has identified the concept autonomy as crucial in the modern discourse on technology and politics meaning self-governing and independence of externally imposed restrictions. Autonomous technology becomes something that excludes human control. Technology and institution are thus sometimes even seen as taking on a life of their own, autonomously threatening humanity almost in a predetermined way.[7] Generally, the different works addressing the modern condition describe a shadow of technological dominance over institutions implying a risk for humanity to be enslaved without control of its own history or future as claimed by system thinkers such as Jacques Ellul.[8]

An explanation of the perceived technological dominance over institutions is framed by the concept of technological imperative describing the situation of technology demanding a restructuring of the institutional contexts in order to be operable.[9] A case could be small-scale renewable power production, for instance, in the form of windmills, which has been proved successful when ownership and residence are linked to each other so that owners are required to live in the vicinity of the physical operation of the wind-powered turbines in order to share both inconvenience and benefit from ownership.[10] As a consequence, proponents of wind energy often support that legislation and subsidies for the construction and operation of windmills are designed to encourage local interest and participation. Although small-scale renewable power production is not the most archetypical illustration of technological imperative, the example nevertheless shows how technology may demand a restructuring of institutions also in cases where its salient features are others than the exploitation of limited resources and the demand for high investments.

Much of the different feelings and sentiments connected to modernity and its features are obviously generated and mediated by technology, not the least by rapid innovation and social change.[11] The German sociologist Ulrich Beck, for instance, defined modernization as "surges of technological rationalization and changes in work and organization" and adds changes of lifestyles, power structures, views of reality and norms of knowledge to mention only a few characteristics.[12] Others, like Marxist Herbert Marcuse, have seen technology as the means of a more general ideology of rationality saturating modernity leading to thought styles and practices dominated by control and a technological imperative forcing an institutional imperative where alternatives are absent.

> The incessant dynamic of technical progress has become permeated with political content, and the Logos of technics has been made into the Logos of continued servitude. The liberating force of technology – the instrumentalization of things – turns into a fetter of liberation; the instrumentalization of man.[13]

It does not seem too far-fetched to contextualize Marcuse's pointing out of coercion and constraint in Western liberal society by reminding how images of freedom in the West were often contrasted to communist political repression during the peak of the cold war. Marcuse seems to have attempted to moderate this view by describing technology as the Western tool of coercion in contrast to the political dictatorships of the East.

Needless to say, such deterministic approaches have been discussed thoroughly. Marcuse's ideas have been coupled to his teacher Martin Heidegger's philosophy of technology and named an essentialist tradition in which determinism can be broken.[14] This essentialist tradition has furthermore been contrasted to constructivism where technology is subservient to institutions or other social spheres.[15] In the ensuing debate, eternal questions of structure, agency and change have dominated, although with the role of technology as the central problem. Is change possible given existing economic institutions such as capitalism or can agency enrol other institutions in order to create new alternatives?[16]

These were problems engaging a number of thinkers during the 1970s, and the answers they supplied pointed towards a theory of relations between technology and institution in a political dimension that leaves quite some space for action. Perhaps they have in common a certain degree of reaction to Marcuse's seemingly inevitable development of technological coercion of the West. One important response was the idea that the oppressive technologies dominating industrialized society was the result of political and economic change that had created capitalists who used technology to exercise power and control.[17] There are interpretations and legitimations of technology, which see it neither as a neutral tool in the hands of an elite, nor as a determinist force that itself transforms society. Instead, technology is the result of social relations and political actions, historically developed and used to dominate others. Technology thus mirrors and strengthens a certain social and political order. In order to come to terms with the historically oppressive technologies, new, alternative ones need to be developed. But this is possible only if a new social and political order has first managed to take power. Only with alternative politics and power is it possible to develop alternative technologies.[18]

These ideas can be put in the context of both Marxism and social constructivism. Marxist ideas resound in the perspective of technology as an expression of political struggle and political struggle as a prerequisite for technological change while simultaneously also expressing the optimism of creating alternatives as expressed in social constructivism. Perhaps these ideas can best be described as political constructivism in terms of social constructivism on a different scale. Implicit are of course ideas of performativity, that theories pointing to the value of agency and political action have a higher potential to contribute to institutional and technological change.

Another idea that has been formulated in the context of modernity and Marxism is the emergence of a new professional class of intellectuals and technical intelligentsia between old moneyed bourgeoisie and public or private bureaucracy. The notion that an identifiable class of professionals has developed a position of either independence or hegemony, or anything in between the two, by claiming expert knowledge of science and technology, has resonated in much of the literature on technological and institutional change from the 1960s.[19] Sometimes this class is seen as following an agenda of its own, other times one of some other ruling class. At any rate, the new class has been seen as being in a privileged position when developing the relations between technology, institutions and change.

The German philosopher Jürgen Habermas built on notions like this to develop a critique of technocracy, government by a group of technical experts and bureaucrats. He showed how elites, by encouraging rational choice as well as instrumental and strategic action, transformed practical problems about the good life to technical problems for experts, thus abolishing the need for public discussion of values. In this way, technology defined as legitimate interest in control of nature becomes an ideology – a system of thoughts, ideas and truth-beliefs that shrouds the value-laden character of decision making maintaining (capitalist) status quo.[20]

Ulrich Beck is nevertheless one of the most prominent thinkers on modern condition using his own concept of risk society to characterize it. In his view, modern institutions are focused on risk management through rational control as a consequence of a number of technologies such as genetic or nuclear.[21] There is a trend, unintentionally inherent in technological change towards automation, to increase risk but simultaneously to become more aware of it. Whether our modern society is more hazardous than earlier ones is thus not of interest. Instead, the key point is that today's institutions are infiltrated by notions of different risks and that they are increasingly active in trying to control them in different ways, for instance, through the institutions of insurance. Risk and reflexivity in the meaning of self-confrontation of risk is of central interest to Beck. Simultaneously, protection from danger decreases as threat increases according to Beck finding a number of examples of organized irresponsibility when individuals or organizations are released from responsibility for causing pollution or other hazards.

Another stance is the stress of both the advantages and the threats characterizing modern life. Marshal Berman has claimed that to be modern is:

> to find ourselves in an environment that promises us adventure, power, joy, growth, transformation of ourselves and the world – and, at the same time, that threatens to destroy everything we have, everything we know, everything we are. Modern environments and experiences cut across all boundaries of geography and ethnicity, of class and nationality, of religion and ideology; in this sense, modernity can be said to unite all mankind. But it is a paradoxical unity, a unity of disunity: it pours us all into a maelstrom of perpetual disintegration and renewal, of struggle and contradiction, of ambiguity and anguish.[22]

These tendencies are perhaps most notable in the context of cognitive artefacts defined as tools of thought and technologies to complement and enhance our cognitive abilities such as writing, reading, calculators and so on, where representations are crucial.[23] Many observers of the influence of cognitive artefacts as well as other technologies claimed to shape human understanding of the world, of each other and of ourselves have simultaneously stressed their powerful mandate to determine individual impressions and thus in extension our view on institutions.

One example is the mirror, which started to become more common in the Western world during the 18th century. According to theories of psychoanalysis, recognition of one's mirror image generates the first and most central splitting of the self, the mirror stage, which leads to an objectification of the personal subject.[24] In this way, the mirror has been viewed as a prerequisite of an observed process of individualization in modern Western society, a process that in turn has had far-reaching consequences for institutions as well.[25] Similarly, the introduction and distribution of glass windows, facilitating both influx of light and the possibilities of uncontrolled gazing into zones of domesticity as well as keeping watch over the surroundings from the inside, may hypothetically have changed how the borders between private and public life shifted in Western Europe from the 17th century.

To sum up, technological and institutional change is crucial for different theories of modernity. Although modernity traditionally has underscored the emancipative force of what seems to have been institutionally distributed growth due to technical change, the modern thinkers of the past decades have to an increasing extent stressed, with Marcuse and Ellul as examples, how complexes of institutions and technologies limit individual and institutional freedom. One reaction has been the connection of emancipation through new technologies with visions of closer links between technologies, institutions and individuals as expressed by Berman. The emergence of a modern "unity of disunity" can of course be explained in many different ways with cognitive artefacts as one example out of many others.

Notes

1. Langdon Winner, *Autonomous Technology: Technics-Out-of-Control as a Theme in Political Thought* (Cambridge, Mass: The MIT Press, 1977), 46–50; Thomas J. Misa, Philip Brey & Andrew Feenberg, eds, *Modernity and Technology* (Cambridge, Mass: The MIT Press, 2003). For a review of the idea of progress from mid-17th century, see: John Bagnell Bury, *The Idea of Progress: An Inquiry into Its Origin and Growth* (London: Macmillan, 1920).
2. Brooks Harvey, "Technology, Evolution and Purpose", *Dædalus* 109:1 (Winter, 1980), 65–81, p. 27.
3. David Harvey, *The Condition of Postmodernity: An Enquiry into the Conditions of Cultural Change* (Cambridge, Mass: Blackwell, 1989), 3–38.
4. Max Horkheimer & Theodor W. Adorno, *Dialektik der Auklärung* (New York: Social Studies Association, 1944).
5. Quoted from: Winner, *Autonomous Technology*, 14.
6. Ibid., 29.
7. Siegried Giedon, *Mechanization Takes Command: A Contribution to Anonymous History* (Oxford: Oxford University Press, 1948).
8. Jacques Ellul, *The Technological Society*, original in French 1954 (New York: Vintage Books, 1964).
9. Winner, *Autonomous Technology*, 100.
10. John Howells, *The Management of Innovation and Technology* (London: SAGE Publications, 2005), 67–71.
11. Helga Nowotny, "Introduction: The Quest for Innovation and Cultures of Technology", in: *Cultures of Technology and the Quest for Innovation*, ed., Helga Nowotny (New York: Berghahn Books, 2006), 1–23.

12 Ulrich Beck, *Risk Society: Towards a New Modernity*, original in German 1986 (London: SAGE Publications, 1992), 50, note 1.
13 Herbert Marcuse, *One-Dimensional Man: Studies in the Ideology of Advanced Industrial Society* (Boston: Beacon Press, 1964), 117.
14 Andrew Feenberg, *Heidegger and Marcuse: The Catastrophe and Redemption of History* (London: Routledge, 2004).
15 Andrew Feenberg, "From Essentialism to Constructivism: Philosophy of Technology at the Crossroads", in: *Technology and the Good Life?*, eds, Eric Higgs, Andrew Light & David Strong (Chicago: The University of Chicago Press, 2000), 294–315; David Edward Tabachnick, "Heidegger's Essentialist Responses to the Challenge of Technology", *Canadian Journal of Political Science* 40:2 (2007), 487–505.
16 Andrew Feenberg, *Between Reason and Experience: Essays in Technology and Modernity* (Cambridge, Mass: The MIT Press, 2010); Tyler Veak, "Whose Technology? Whose Modernity? Questioning Feenberg's *Questioning Technology*", *Science, Technology, & Human Values* 25:2 (2000), 226–237; Andrew Feenberg, "Do We Need a Critical Theory of Technology? Reply to Tyler Veak", *Science, Technology, & Human Values* 25:2 (2000), 238–242.
17 David Dickson, *Alternative Technology and the Politics of Technical Change* (London: Fontana Books, 1974).
18 Others have insisted on the need to first develop alternative technologies as a means to transform political life and power relations, see: Robin Clarke, "The New Utopias", *New Scientist* 62:898 (May 16, 1974), 423; Adrian Smith, "The Alternative Technology Movement: An Analysis of Its Framing and Negotiation of Technology Development", *Human Ecology Review* 12:2 (2005), 106–119.
19 Jürgen Habermas, *Toward a Rational Society: Student Protest, Science and Politics*, original in German in 1968 and 1969 (Boston: Beacon Press, 1970).
20 Jürgen Habermas, *Theory and Practice*, original in German in 1963, 1966 and 1968 (Boston: Beacon Press, 1973).
21 Beck, *Risk Society*; Anthony Elliott, "Beck's Sociology of Risk: A Critical Assessment", *Sociology* 36:2 (2002), 293–315.
22 Marshall Berman, *All That Is Solid Melts into Air: The Experience of Modernity*, original in 1982 (New York: Penguin Books, 1988), 15.
23 Donald A. Norman, *Things that Make Us Smart: Defending Human Attributes in the Age of the Machine* (Boston: Addison-Wesley, 1993); Jill Walker Rettberg, *Seeing Ourselves through Technology: How We Use Selfies, Blogs and Wearable Devices to See and Shape Ourselves* (Basingstoke: Palgrave Macmillan, 2014).
24 David Ross Fryer, *The Intervention of the Other: Ethical Subjectivity in Levinas and Lacan* (New York: Other Press, 2004).
25 Alan Macfarlane & Martin Gerry, *Glass: A World History* (Chicago: The University of Chicago Press, 2002).

Further reading

Misa, Thomas J., Philip Brey & Andrew Feenberg, eds (2003), *Modernity and Technology* (Cambridge, Mass: The MIT Press).

Winner, Langdon (1977), *Autonomous Technology: Technics-Out-of-Control as a Theme in Political Thought* (Cambridge, Mass: The MIT Press).

13
Postmodernity

Abstract: *Postmodern thinking may be derived from cybernetics and systems theory, where ideas of hybridity between technological and social thinking are common. Views on the merging of technologies and institutions thus to some extent rely on reflexivity. One consequence has been hopes for emancipation through new technologies such as computers for communication and information management leading to visions of closer links between technologies, institutions and individuals. These visions have resulted in postmodern experiences of increased heterogeneity as well as a reaction to modernity, for instance, in terms of social acceleration and fractalization.*

Keywords: acceleration; cyborg; fractalization; hybridization; information system; postmodernity; space compression

Kaiserfeld, Thomas. *Beyond Innovation: Technology, Institution and Change as Categories for Social Analysis*. Basingstoke: Palgrave Macmillan, 2015. DOI: 10.1057/9781137547125.0015.

Speculations about technological influence on our perception of the world and each other made ideas about steadily increased technological efficiency seem less reasonable during the 20th century. As a result, grand narratives were doubted, as was the very idea of one single coherent interpretation of nature and ethics. An early observer of these tendencies was Jean-Françios Lyotard.[1] His point of departure was the insight that many of the leading sciences and technologies from the 1950s and onwards were connected to language and linguistics. Especially, computer technologies and computer science could be expected to have a considerable impact on knowledge and its management.

The idea that computers and new information management would revolutionize our understanding of relations between technology and institution can be traced back to a number of cold war discourses.[2] Important was, for instance, the development of computer power strong enough to model military and political problems, a development, which, in turn generated further increased computer power and software developments leading to ideas of computers' recursive self-improvement in the 1960s.[3] Starting out with concrete problems of calculating trajectories with the help of early digital computers in the 1940s, cold war nuclear strategies and policies followed as an important consequence in turn informing the design of computers and their programs.[4]

Information management technologies became the basis for notions of a new age where technologies, minds and institutions would merge, not the least within the discourse of cybernetics, the study of information feedback systems and communication statistics, during the 1950s and 1960s.[5] In a vision of information cut loose from hardware, machines would take over many different human capabilities and cybernetic organisms, cyborgs for short, would evolve with powers surpassing those of both humans and machines creating a posthuman era.[6] In popular culture, computers mimicking neurological processes were abundant as in Stanley Kubrick's enigmatic feature *2001: A Space Odyssey* from 1968 where the computer HAL 9000 got a nervous breakdown due to contradictory instructions. Within this framework, advanced communication with globalization as a consequence was part of the dream together with visions of fusions between humans and machines in notions clearly forecasting ideas of hybridity to be dealt with in the next chapter.

The rise of information technologies thus had important consequences for ideas about the relations between technologies and institutions in postmodern culture. Exponents of these trends were, for instance,

Austrian biologist, Ludwig von Bertalanffy, who had theorized about biological systems since the 1930s and developed his thoughts into a general systems theory after World War II, and American mathematician Norbert Wiener who had entered the field from military applications.[7] Their ideas and others' were brought into social thinking with speculations of far-reaching consequences. In return, cybernetics and information technologies also had consequences for biomedicine when computers were shrunk to fit the demand of individual biologists.[8] Lyotard's interpretation was that these new technologies had shifted the conditions for describing history as well as the purpose of agency:

> The grand narrative has lost its credibility, regardless of what mode of unification it uses, regardless of whether it is a speculative narrative or a narrative of emancipation. The decline of narrative can be seen as an effect of the blossoming of techniques and technologies since the Second World War, which has shifted emphasis from the ends of action to its means.[9]

It was the combination of engineering feats and social thinking that made the ideas of information systems exert such a strong influence in such a wide range of different technologies, from automated precision warfare to civilian cell-phone communication and use, having effects also for ideas about the purpose of institutions. World War II functioned as a catalyst for a transformation of the goal of state research from communal welfare to military dominance.[10] After 50 years of technological and institutional change along these visions of automated information management emanating from military contexts, it is perhaps not surprising that modernism had transformed into notions of postmodernism and beyond. Observers have at least claimed that modernity relied on science as postmodernism relied on technology.[11]

The increasing multitude of voices in postmodern society implied a critique of grand narratives as well as grand theories regarding social change including ideas of general relations between technologies and institutions and their dynamics. This multitude thus had consequences for views on institutional and technological change. Interestingly enough, however, Lyotard's analysis to a large extent relied on a grand narrative itself.

Another important critic of modernity has been sociologist Bruno Latour who has viewed it as built upon the division between nature and society.[12] The simultaneous division between nature and society has, however, been accompanied by a constant hybridization of humans and non-humans. According to Latour, these two processes of purifying nature and society through division while simultaneously constantly mediating and translating between nature and society through the creation of hybrids in our lives, from walking canes to clones, characterize modern thinking and practices. Although technology has been used for centuries to enhance our senses, for instance, through the invention of eyeglasses during medieval times, the general notion today, right or wrong, is that these tendencies have been amplified considerably over the past decades. Moreover, it seems as if present-day institutions encourage such developments in a vast array of areas, from prenatal testing to Google Glass, a wearable computer with an optical head-mounted display. The result has been the acceleration of creation of hybrids of nature and society in the long term undermining the division.

Another way to analyse interactions between technology and institution is to acknowledge the rejection of grand narratives and instead depart from micro-level case studies. This has indeed been the background to Misa's contention that studies of dynamics in greater detail and over shorter time periods tend to highlight institutional factors as exogenous, an observation touched upon in Chapter 11. The result has been an increased interest in relativistic perspectives and stress on contexts while playing down the autonomy of technology as often expressed in modernistic accounts.

Over the past decade or two, there has, however, been a revival for broader time perspectives, for instance, accompanied by analyses along the lines of new institutionalism described in Chapter 9. In addition, intensified inter-cultural relations as mirrored in the concept of hybridization, that is, ways in which forms are separated and recombined to constitute emerging new practices, have also been used as a framework to understand the interplay between technology and institution. Moreover, mobility, migration and multiculturalism have been pointed out as processes behind compression of time and space with intensified hybridization as a consequence, in turn of course, to a large extent dependent on changing conditions created by cheaper and more efficient transport and communication technologies.

One of the most important thinkers about compression of time and space is social theorist and Marxist David Harvey. He has stressed how time and space compression has resulted from more than mobility alone. In a world where capitalist institutions have transgressed national borders to such an extent that an American tourist may use a Thai automatic teller machine to make a withdrawal from a German bank with head office in the United Kingdom or even pay for goods purchased from a Chinese firm, this is a typical example of transnational space compression and interaction, which have had consequences for how we think about institutional and technological change. Twenty years later, we have learnt that money transfer led the way for communication through social media or peer-to-peer technologies such as Skype as if there were no borders at all, no institutions to balance between technological change and institutional inhibition as mentioned in Chapter 2.

One very important backdrop to this development is the idea that the present view of speed and space has fundamentally transformed modern institutions. One of the most important heralds of this line of thought is Paul Virilio who has explained how acceleration lies at the heart of today's institutions. Technological change promoting acceleration and control of space as well as temporal structures in general is primarily driven by war efforts. The by far most important institution is thus the military and the military-industrial complex, and its most important feature is the war of movement carried by transportable and accelerated weapons systems. All other institutions are to a larger or smaller extent influenced by war and the military.[13] This seemingly institutional determinism is outweighed by a technological determinism, for instance, shown in the interest in technological accidents that is seen as an unavoidable consequence of any given technology such as the ship or the airplane.[14] More recently, Virilio has also stressed the military reliance on information and intelligence as well as technologies for remote control leading to "fractalization" of physical space with repercussions for the restructure of civil institutions and space.[15]

Comparable patterns of analysis, although not putting the same stress on the military institution for the importance of acceleration, are represented by Hartmut Rosa's ideas about social acceleration characterizing modernity and change in general in our society.[16] Since industrialization, Western society has to a large extent been able to shape its conditions just as its individuals have been able to shape their lives. But the past two or three decades have seen a change. According to Rosa, personal and collective autonomy is no longer as obvious as it once was and this is explained by a number of different acceleration processes that have started to limit individual opportunities rather than expand them as they used to.

Rosa distinguishes three different forms of acceleration. First, the technical acceleration of production, transport, communication and information management has been going on since early industrialization through improvements of existing technologies and the introduction of new ones. The technical acceleration has resulted in acceleration of transportation and changed spatial relations; in acceleration of communication and changed social relations; and in acceleration of production and changed relations to things. All of these have influenced our relations to time.[17] The engine for this acceleration is profit. Secondly, there is also the acceleration of social change in terms of ideological outlook and value systems mirrored in change of jobs, parties and partners as well as the relation between different groups. The engine here is professionalization and specialization. Thirdly, the acceleration of the pace of life is felt in pressure not to miss opportunities and a scarcity of time despite technical acceleration supplying more leisure. The engine behind this acceleration of pace of life is secularization since religion's promise of eternal life (implying that actual lifespan is unrelated to experiences of the good life) has been substituted for physical experiences where as much as possible is to be fitted into each individual's lifespan. All three forms of acceleration strengthen one another.

Although the thoughts of Virilio and Rosa may not seem as original, they both have the advantage of putting time compression and acceleration into the context of modernity, although from different perspectives. In common, they have a stress on Western perspectives and moreover on middle-class experiences. They both can be contrasted to David Harvey's stress on time and space compression, which in his version have consequences for transnational interaction. Taken together, they all point to conditions crucial for the understanding of technological and institutional change of the modern age. What they all underscore is neither technology nor institution, but instead the rate of change and how it influences relations between technologies and institutions beyond modernity whether in terms of fractalization, social acceleration or transnational interaction. In the case of Virilio, it is clear how specific institutions such as the military may effect other institutions with technologies mediating the influence. In Rosa's case, the acceleration of technological change is intimately related to the rate of institutional change with both prompting each other. In common, they also have their pessimistic conclusions regarding the future of modern life as well as the stress of change rate for the understanding of today's relations between technology and institution.

These visions have resulted in postmodern experiences of increased heterogeneity as well as new understandings of modernity. Whether such visions can be called deterministic or not, they may be interpreted as expansions of Moore's law regarding quantitative measures of technological change to qualitative analyses of institutional change. Just as postmodern thought can be related to the cybernetics movement of the 1950s and 1960s, notions of fractalization, social acceleration or transnational interaction can be viewed in the perspective of increased digitalization.

Notes

1 Jean-François Lyotard, *The Postmodern Condition: A Report on Knowledge*, Theory and History of Literature, Vol. 10, original in French 1979 (Manchester: Manchester University Press, 1984), xxiv.
2 N. Katherine Hayles, *How We Became Posthuman: Virtual Bodies in Cybernetics, Literature, and Informatics* (Chicago: The University of Chicago Press, 1999).
3 Terrell Ward Bynum, "Norbert Wiener and the Rise of Information Ethics", in: *Information Technology and Moral Philosophy*, eds, Jeroen van den Hoven & John Weckert (Cambridge: Cambridge University Press, 2008), 8–25.
4 Paul N. Edwards, *The Closed World: Computers and the Politics of Discourse in Cold War America* (Cambridge, Mass: The MIT Press, 1996).
5 Ronald R. Kline, "Cybernetics, Management Science and Technology Policy: The Emergence of Information Technology as a Keyword, 1948–1985", *Technology and Culture* 47:3 (2006), 513–535.
6 Terrell Ward Bynum, "Ethical Challenges to Citizens of 'The Automatic Age': Norbert Wiener of the Information Society", *Journal of Information, Communication and Ethics in Society* 3:2 (2004), 65–74; Terrell Ward Bynum, "Philosophy in the Information Age", *Metaphilosophy* 41:3 (2010), 420–442.
7 Manfred Drack, "Ludwig von Bertalanffy's Early System Approach", *Behavioral Science* 26:5 (2009), 563–572.
8 Joseph November, *Biomedical Computing: Digitizing Life in the United States* (Baltimore: The Johns Hopkins University Press, 2012).
9 Lyotard, *The Postmodern Condition*, 37–38.
10 Philip Mirowski, "The Scientific Dimensions of Social Knowledge and Their Distant Echoes in 20th-Century American Philosophy of Science", *Studies in History and Philosophy of Science* 35:2 (2004), 283–326; Helen Longino, "Whither Philosophy of Science?", *Studies in History and Philosophy of Science* 36:4 (2005), 774–778; Philip Mirowski, "Hoedown at the OK Corral: More Reflections on the 'Social' in Current Philosophy of Science", *Studies in History and Philosophy of Science* 36:4 (2005), 790–800.

11 Paul Forman, "The Primacy of Science in Modernity, of Technology in Postmodernity, and of Ideology in the History of Technology", *History and Technology* 23:1 (2007), 1–152. For a review of the debate that followed, see: Jennifer Karns Alexander, "Thinking Again about Science in Technology", *Isis* 103:3 (2012), 518–526.
12 Bruno Latour, *We Have Never Been Modern*, original in French 1991 (Cambridge, Mass: Harvard University Press, 1993). See also Mark Elam, "Living Dangerously with Bruno Latour in a Hybrid World", *Theory, Culture & Society* 16:4 (1999), 1–24.
13 Paul Virilio, *Speed and Politics*, original in French 1977 (Los Angeles: Semiotext(e), 2006).
14 Paul Virilio, *The Politics of the Very Worst*, ed., Sylvère Lotringer (Los Angeles: Semiotext(e), 1999).
15 Paul Virilio, *The Information Bomb*, original in French 1998 (London: Verso, 2000).
16 Hartmut Rosa, *Social Acceleration: A New Theory of Modernity*, original in German 2005 (New York: Columbia University Press, 2013).
17 Ibid., 97–107.

Further reading

Latour, Bruno (1993), *We Have Never Been Modern*, original in French 1991 (Cambridge, Mass: Harvard University Press, 1993).

14
Hybridity and Technology Transfer

Abstract: *Defining the concept of hybridization as ways in which forms become separated from existing practices and recombined with new forms in new practices highlight the problematic division between natural and artificial. When further analysing our technological contexts, a conclusion is that relations between institutions and technologies should be characterized more by technological stability than institutional. Moreover, technology transfer from one institutional context to another and the subsequent adjustments of both technologies and institutions is a common feature of today as well as the specific case of generic technologies. Analyses of the past two decades stress the mixing of technologies and institutions over cultures and geographical distances when highlighting what are judged to be important components of present changes.*

Keywords: creole technologies; generic technologies; glocal technologies; hybridization; militainment; posthumanism; technological transfer; transhumanism

Kaiserfeld, Thomas. *Beyond Innovation: Technology, Institution and Change as Categories for Social Analysis*. Basingstoke: Palgrave Macmillan, 2015. DOI: 10.1057/9781137547125.0016.

Another strand of the cybernetic movement, the notion of the posthuman hybrid bioengineered-techno-body reviewed earlier in Chapter 7 on agency and in the previous chapter in connection to cyborgs, has equally important consequences for our understanding of relations between technology, institution and change. The key concept here is hybridity, defined as the combination of "the human and the nonhuman, the technical and the social".[1] Mikael Hård and Andrew Jamison have characterized the concept by pointing to certain aspects of human interaction:

> In order to make use of our knowledge and artifacts, we not only have combined human and nonhuman entities, but also brought together previously separate social roles and identities, with their different skills and competencies, into new hybrid forms.[2]

The master of hybridity is Bruno Latour, who has discussed the concept thoroughly throughout his writings. Examples of this perspective is everywhere around us. The clothes on our bodies are made of natural fibres such as cotton or wool or artificial fibres such as nylon. But this division between natural and artificial is in itself problematic to uphold if scrutinized more closely. Wool and cotton fibres originate from animals and plants respectively, but need to be processed thoroughly and in many different steps – one might describe it as appropriated – in order to make it suitable for cloth weaving and production of clothes. Nylon, developed in the 1930s as a substitute for silk, and other so-called artificial fibres are petroleum products thus originating from fossil material that has been processed in a number of steps as well.[3] In fact, both types of fibres have origins in nature, but need to be processed thoroughly to fit as fibres for clothes. Most of our clothes today consist of a mix of fibres extracted from both fossil and live biological sources. This processing of natural material from animals, plants or fossils are all examples of translation processes creating hybrids of society and nature as well as of technology and institution. A garment is technology and institution seamlessly interwoven and ready to wear.

This is also one of the main points made by the proponents of transhumanism, a movement trying to create technologies and institutions to facilitate the convergence of nano- and biotechnologies together with information technologies, cognitive sciences and future means in order to enhance and strengthen human capabilities such as cognitive skills as well as prolonging life expectancy or even use cryogenics to be able to revive ceased humans. This, if anything, could rightly be called posthumanism. The idea is to use technologies to improve individual's bodies in order to fight poverty, disease and disability. Some claim this is not only an option but that there is an ethical imperative to enter the transhuman phase where humans can take control of their own evolution. Different lines of arguments have been used to promote these notions including ethical and philosophical perspectives on the human condition as well as deterministic ideas of a more or less unavoidable process leading to the irreversible transformation of human life.[4]

From more conventional perspectives, institutions are instrumental in facilitating hybrid combinations. For instance, technologies such as gender reassignment surgery are strongly related to an increase in transgendered relations and new patterns of family formation, obviously to a large extent relying on legal frameworks as well as the cover of social security and different norms in different social settings. Other examples are the development of so-called smart grids to facilitate better matchmaking between energy production and consumption, not only through differentiated pricing, but also through the possibility for power consumers to pool power consumption and demand specific production methods. In these cases, the engagement of technology and institution cannot be separated analytically, indeed the foundation for applying the concept of hybridization.[5]

Over the past decades, there has thus been a resurge in the literature on technological and institutional amalgamations departing from the notion that both technologies and institutions are becoming ever more important for interaction across boundaries, not only those constituting nation states, cultures or time and space, but also between different activities such as artistry or engineering. The growing contacts and mutual influences have also induced postmodernist critique, especially incredulity toward grand narratives such as Marxism, Nationalism or any type of general explanation of technological and institutional change. It is nevertheless the broader systems of thought departing from Marxism via modernism and postmodernism to more recent system observations that will be invoked in order to further discuss ideas of technological and institutional change.

More specific are the observations often made today that there seem to be intensified inter-cultural relations. Using the concept of hybridization, meaning the ways in which forms become separated from existing practices and recombined with new forms in new practices, the developments have thus been described in a recently published encyclopaedia:

> Hybridity has become ordinary and a part of everyday life in a world of intensive intercultural communication, multiculturalism, growing migration and diaspora lives, and eroding boundaries, at least in some spheres. Hence, hybridity has become a prominent theme in cultural studies. The emergence of new hybrid forms and practices indicates profound changes that are taking place as a consequence of mobility, migration, and multiculturalism. However, hybridity thinking also concerns already existing or, so to speak, old hybridity, and thus involves different ways of looking at historical and existing cultural and institutional arrangements.[6]

Here, mobility, migration and multiculturalism are pointed out as factors behind recombination of new practices. To be sure, this notion is not entirely new. In a world of increasing transport and communication relying on expanding technological systems as well as institutions making these systems work on regional and global scales, similar ideas have been presented repeatedly. One prominent example where contacts between different civilizations are stressed for the promotion of inventive activities regarding institutions as much as technologies is the theory from the 1960s of the development of civilizations through interaction between them.[7] This should be contrasted to earlier historians' view on a struggle between civilizations where, for instance, a "creative minority" finds solutions to challenges posed by environment or threats from other civilizations.[8] Such broad cultural theories can also be combined with ideas of how competence in different areas affects individual creativity as implied in the concept of technological frames mentioned earlier in Chapter 6.

The description of our present world as shaped by mobility, migration and multiculturalism together with normative claims that integration of and interaction between different areas of expertise is the most efficient way to promote new and better practices in different domains such as technology brings us to a rather optimistic view on future potentials. In innovation studies and other areas where creativity and inventive capabilities are analysed, this is nothing new. However, while a lot of analytical interest is put on communication, information and medical technologies, theoretical frameworks for analysing institutional and technological change regarding other, older technologies must not be overlooked in this context.

Historian of technology David Edgerton reminds us how our view on technologies and institutions is tainted by the constant stress on invention, innovation and diffusion as well as an iteration of path-breaking technologies, in the 20th century primarily flight, nuclear power, contraception and the Internet.[9] When analysing our technological context, he has found old technologies dominating the scene while most of us image the world infused with the latest and most advanced. Here is another alternative narrative challenging common knowledge. The Internet is not new, but was launched in 1965 and is thus commemorating its semi-centennial anniversary this year. But nobody will celebrate it simply because our image of the Internet is not that of an old and worn technology, but that of something new and ever changing that has not yet found its institutional mould. Likewise, nuclear power, emanating from atomic weapons programmes, is more than 50 years old today. In addition, and as Edgerton points out, nuclear power has not been lucrative or helped solving an increasing demand for electric power. Instead, the early so-called commercial nuclear reactors were built to produce weapons-grade plutonium with electrical power as a spin-off. When calculating the costs for nuclear power, this has proven an extremely expensive technology, presumably drawing resources from alternative cheaper and more efficient power-producing technologies.

In this way, Edgerton shows how much of modern technology is in fact old or inefficient, or both. The relations between institutions and technologies should be characterized more by technological stability than institutional. This does not imply that technology does not play an important role, however, but a different one.

> We should be aware, for example, that most change is taking place by the transfer of techniques from place to place. The scope for such change is enormous given the level of inequalities that exist with respect to technology. Even among rich countries there are very important differences in, for example, carbon intensity. If the USA were to reduce its energy-use levels to those of Japan, the impact on total energy use would be very significant. But for poor countries, as well as for rich ones, such a message is often unwelcome. For *imitating* is seen as a much less worthy activity than innovating. To imitate, to replicate, is to deny one's creativity, to impose upon oneself what was designed for others by others. "*Que inventen ellos*" ("Let *them* invent") is seen not as sensible policy advice, but a recipe for national humiliation. To have technology or science is, it is often deeply felt, to create something new.[10]

This stress on technological transfer is not novel. Instead, scholars have pointed out how geographical technology transfer seldom occurs wholesale from one regional and cultural setting to another, but entails a more complex process including, for instance, inventive responses from recipients.[11]

This is even more often the case when dealing with what Edgerton has named creole technologies "which finds a distinctive set of uses outside the time and place where it was first used on a significant scale".[12] Transferred technologies are a subset of creole ones and perhaps the simplest version. But creole technologies may also form hybrids with existing local technologies. This process has also been pointed out, however, as a problematic one. In his grand ecological narrative, Jonathan Radkau has described the increasing technology transfer as a threat:

> As the world becomes more tightly knit together, the transfer of knowledge and technology extends over increasing greater distances. Technologies become detached from the environment in which they arose and to which they were adapted, and this creates a new kind of environmental risk. The heavy plow is used on soils for which it was not created and which it exposes to erosion. Water-intensive technologies that stem from water-rich Western and Central Europe are transferred to water-poor regions of the world. Of course, such technology transfer is not always blind to the regional conditions: the history of technology is not only a history of the diffusion of technology, but also of its adaptation. That history, however, is messier: it is more a history of details than of great technological ideas, which is why only its rudiments have been written to date.[13]

Radkau's solution to the threat of transfer is adaptation to local conditions, what in Edgerton's version is named creole technologies. Apparently, the force behind applying technologies developed elsewhere by reworking or editing them to fit new local or regional contexts and conditions is pointed out as a road ahead by more than one commentator. Both these cases can easily be combined with a general notion of a development of technologies and institutions that favour commonly managed use values rather than privately generated exchange values to frame the issue in Marxian terminology. A third version of an essentially similar notion of technological change as dependent on local developments is framed in the term localization to describe how local intermediaries modify technology to fit local geographical, meteorological and institutional conditions.[14] In these cases, the strength is the multitude of alternatives and contexts that can come into play when looking for common ground in both problems and solutions.

But technologies may very well be creole without having to travel geographically at all. Take simple examples from everyday life of any disabled in Western society. Speech impediments may lead to intimate reliance on computers for synthesized speech originally developed for other purposes. Such solutions may both strengthen and weaken the institutional frameworks of disability in which they are supported and developed.[15] Another common example is genetic technologies, originally developed to improve efficiency in plant breeding and animal husbandry. Over the past decades, genetic technologies have been creolized through application in prenatal testing, which together with an individualization of health services has led to a response in terms of institutional mobilization and identity formation among social groups threatened by the new possibilities for individual prenatal testing.[16] Both these cases are examples of technology transfer that is not geographical, but functional, a type of technology transfer also named generic technologies. A third example, and perhaps most defying, is the use of entertainment computer games for practicing combat situations such as the manoeuvring of UAVs. In this case, computer software developed for entertainment has achieved such a level of accomplishment that it is also used for military training. In fact, the defence industry has developed software to be used as entertainment and practice simultaneously, which is called militainment.[17]

As used in the literature, hybridity and creolization frame the notion that technology is becoming ever more intertwined with our institutions as well as hybridized with our bodies and minds. They are to some extent framed in the notion of generic technologies defined as "technology the development of which will have implications across a range of other technologies" or general-purpose technologies.[18] Generic technologies are to some extent the opposite of creole since they are developed to be used in a wide range of times and places. Often, however, it seems to imply hybridity. In all cases, they can be described as technologies that are the object of functional transfer and that frequently enough technologies become generic only long after they have been developed for a specific purpose rather than being originally designed for general purposes.

Quite clearly, analyses of the past two decades stress the mixing of technologies and institutions over cultures and geographical distances when highlighting what are judged to be important components of present change. Both institutions and technologies seem to be less local and more global, at least according to contemporary theories about their interaction. The use of concepts such as hybridity and creolization both bear witness to this, as does the hybrid concept of glocal technologies.

We are fast discovering just how interconnected the universe is, and just how interdependent living and non-living systems and sub-systems are upon each other. [–] Humanity urgently needs integral, holistic, cross-disciplinary intellectual conversations and actions to create solutions that address the multidimensionality of the systems with which we must now engage. It is perhaps surprising that, in the face of this growing awareness, modern science and engineering and the agencies which fund them remain entrenched in their traditional silos of activity. In the West, most of these activities are organised so as to resolve short-term, commercial concerns within the narrow agenda, that is, neo-liberalism. Our myopia will be judged harshly by our children's children's children. According to any reasonable moral framework, this will no longer do.[19]

Notes

1. Mikael Hård & Andrew Jamison, *Hubris and Hybrids: A Cultural History of Technology and Science* (New York: Routledge, 2005), xiii.
2. Ibid., 5.
3. M. S. Vassilou, *Historical Dictionary of the Petroleum Industry*, Historical Dictionaries of Professions and Industries 4 (Plymouth: Scarecrow Press, 2009), 390–391.
4. Steve Fuller, *Humanity 2.0: What It Means to Be Human Past, Present and Future* (Basingstoke: Palgrave Macmillan, 2011); Ray Kurzweil, *The Singularity Is Near: When Humans Transcend Biology* (New York: Penguin, 2005).
5. Bruno Latour, *We Have Never Been Modern*, original in French 1991 (Cambridge, Mass: Harvard University Press).
6. Jan Nederveen Pieterse, "Cultural Hybridity", in: *Encyclopedia of Global Studies*, eds, Helmut K. Anheler & Mark Juergensmeyer, vol. 1 (London: SAGE Publications, 2012). Definition of hybridity in: William Rowe & Vivian Schelling, *Memory and Modernity: Popular Culture in Latin America* (London: Verso, 1991), 231.
7. William H. McNeill, *The Rise of the West: A History of the Human Community* (Chicago: The University of Chicago Press, 1963).
8. Arnold J. Toynbee, *A Study of Civilizations*, 12 vols. (Oxford: Oxford University Press, 1934–1961).
9. David Edgerton, *The Shock of the Old: Technology and Global History since 1900* (London: Profile Books, 2006).
10. Ibid., 209.

11 Arnold Pacey, *Technology in World Civilization: A Thousand-Year History* (Cambridge, Mass: The MIT Press, 1990), 50–51.
12 David Edgerton, "Creole Technologies and Global Histories: Rethinking How Things Travel in Space and Time", *Journal of History of Science and Technology* 1 (2007), accessed at: http://johost.eu/vol1_summer_2007/vol1_de.htm, February 17, 2014, see p. 23.
13 Jonathan Radkau, *Nature and Power: A Global History of the Environment*, original in German 2002 (Cambridge: Cambridge University Press, 2008), 11.
14 José Guadalupe Ortega, "Machines, Modernity, and Sugar: The Greater Caribbean in a Global Context, 1812–50", *Journal of Global History* 9:1 (2014), 1–25.
15 Ingunn Moser, "Against Normalisation: Subverting Norms of Ability and Disability", *Science as Culture* 9:2 (2000), 201–240.
16 Christian Munthe, "A New Ethical Landscape of Prenatal Testing: Individualising Choice to Serve Autonomy and Promote Public Health. A Radical Proposal", Presentation at: Individualized Choice: A New Approach to Reproductive Autonomy in Prenatal Screening, Brocher Foundation, Geneva, April 4–5, 2013; Dorothy Roberts & Sujatha Jesudason, "Movement Intersectionality: The Case of Race, Gender, Disability, and Genetic Technologies", *Du Bois Review* 10:2 (2013), 313–328.
17 Roger Stahl, *Militainment Inc. War, Media, and Popular Culture* (New York: Routledge, 2010).
18 Michael Keenan, "Identifying Emerging Generic Technologies at the National Level: The UK Experience", *Journal of Forecasting* 22:2–3 (2003), 129–160, p. 132.
19 Larry Stapleton, "Technology, Culture and International Stability", *AI & Society* 29:2 (2014), 139–142, p. 139.

Further reading

Edgerton, David (2007), "Creole Technologies and Global Histories: Rethinking How Things Travel in Space and Time", *Journal of History of Science and Technology* 1 (2007), available at: http://johost.eu/vol1_summer_2007/vol1_de.htm, February 17, 2014.
Pacey, Arnold (1990), *Technology in World Civilization: A Thousand-Year History* (Cambridge, Mass: The MIT Press).

15
Conclusions

Abstract: *A general conclusion of this book is that existing relations between technology and institution to a large extent determine the theories that are developed in order to describe and understand them. Consequentially, key features identified as necessary to adjust in order to influence and even change the dynamics of institutions and technologies are specific to a certain context set in time and place. Thus there seems to be an ever more positive feedback loop between existing relations, the study of relations, the forming of relations and back to the existing relations again. It is this observation together with today's heterogeneous relations between technology, institution and change that give a review of the present multitude of theories beyond innovation its greatest value.*

Keywords: change; institution; technology

Kaiserfeld, Thomas. *Beyond Innovation: Technology, Institution and Change as Categories for Social Analysis*. Basingstoke: Palgrave Macmillan, 2015. DOI: 10.1057/9781137547125.0017.

A general conclusion from this review is that existing relations between technology and institution to a large extent determine the theories that are developed in order to describe and understand them. Consequentially, key features identified as necessary to adjust in order to influence and even change the dynamics of institutions and technologies are specific to a certain context set in time and place. Marxism followed from industrialism as postmodernism and network theories were a reaction to the cybernetics of the information age.[1] Innovation studies are no exception and should be seen as the heir of liberal consumption markets where successful commercialization of new products was a safe road to economic growth with the entrepreneur as hero. The last, but surely not final, attempt for an Alexandrian solution to cut the Gordian knot of technology, institution and change is perhaps ideas of globalization and hybridity mirroring the capacity for people and artefacts to travel and communicate.

In the introduction of this book, the example of how to manage spent nuclear fuel from nuclear power plants was described to focus on technical innovation while paying much less attention to the development of suitable social institutions to preserve and maintain knowledge and memory about deposits for spent fuel. This book has been written with the intention to review a number of different perspectives on how to understand technological and institutional change of which some could most likely be used to find solutions to this problem. For instance, a neoclassical market could be set up for the trading of rights to produce spent nuclear fuel just as the European Union Emission Trading Scheme has permitted the trade of emission rights for carbon dioxide. This is not, however, a very efficient solution taking into account that nuclear technologies are dual-use technologies where military applications of civil technologies and their products such as spent nuclear fuel is a possibility and goes against different non-proliferation treaties.

Another way of handling the problem, and more along the lines of the reasoning at the end of the preceding chapter and perhaps a much better mirror of today's relations between technology and institution, would be to try to develop a global deposit for all spent nuclear fuel in the world after having pursued a global siting process.[2] Needless to say, such a suggestion is extremely controversial from a number of perspectives as well as contrary to a number of existing international conventions and agreements. Most importantly, it may encourage poorer regions to volunteer as a global deposit in order for rulers or majorities, or both,

130   *Beyond Innovation*

to exchange future land use for present growth and potential personal wealth.

In this way, resources from different countries exploiting nuclear power could nevertheless be funnelled to one single site solving a lot of problems in countries where it is hard to find a suitable site, for instance, Japan. The challenge is to allow for hybridity and creolization, in other words, keeping the solution global while simultaneously applying local solutions in order to avoid isomorphism, which may cause conflict. This can of course be achieved only by involving the very people affected by the construction of the deposit making their consent a necessary, though not sufficient, condition. Likewise, the technological design of the deposit as well as the social institutions developed in tandem must be a concern for all those immediately influenced by its construction. This does not only include discussions on eventual markers signalling the content and position of the site as well as the condition of the spent nuclear fuel deposited in it or institutions initiated to pass on the memory of the construction across generations, but also the very technologies used in order to avoid leaks or human intrusion, for instance, monitoring devices and global distribution of the information obtained through them.

Only by closely linking technology and institution in a coordinated global and local innovation process would it be possible to find a country or region willing to accept the storage of spent nuclear fuel on a global scale and handle the ethical issues arising from creating a common dustbin for nuclear waste. In addition, a global solution is feasible only if all nations can agree on how much spent nuclear fuel each may deposit and at what cost. Just as such a site would be a symbol for waste management on a scale never experienced before, it would also potentially be a symbol for self-sacrifice on a global scale. Most importantly, it would be a solution that affirms the present global interconnectivity as well as the need to address challenges through international cooperation also taking into account local traditions in terms of both technologies and institutions. What are the traditional relations to the underground and its exploitation? Which institutions and technologies have been involved? How can they be used and developed in order to converge with present ideas of how to manage nuclear waste?

Such an effort would certainly neither be an easy undertaking, nor satisfy the expectations generated by the innovation monomania described in the introduction as ruling different policy initiatives on

Conclusions 131

a global scale. Whether it disqualifies as an innovation is of course a matter of definition. Taking into account the expansion of the innovation concept over the past decades, for instance, by the introduction of social innovation, it seems hard to disqualify any new ideas as innovations. The notion to cooperate globally to develop storage for spent nuclear fuel nevertheless diverges from a number of features often connected to innovations and the policy of supporting them in order to achieve economic growth, a notion at the core of innovation studies. First of all, it is not likely to be relevant on any market or lead to economic growth. Instead, the purpose is to manage a negative externality in a responsible way. Secondly, it may, but it does not have to, include any novel technologies since technologies being developed in a number of different countries with commercial nuclear power could work just as well on a global scale although of course depending on the type of bedrock chosen for the deposit. However, the focus is more likely to be on institutional innovation and solution, especially developing new safeguards allowing for a global deposit while still satisfying demands from different nations and international organizations. Thirdly, it necessitates cooperation between a number of stakeholders, nations and international organizations rather than competition. Fourthly, this solution is unique and to be constructed only once. Thus, a major disadvantage with such a global deposit of spent nuclear fuel is that the knowledge accumulated during its development and construction will not be exploited later on as it would have been if a number of different national deposits had been built. Being the first and last of its kind, a global deposit is likely to be as expensive and poorly constructed as any beta version. The savings and improvements usually made when building later versions on the experiences gained from predecessors will be lost.

Of course, this scenario is extremely spectacular and anything but realistic today. But so is Generation IV reactor prototypes calculated to cost close to 11 billion euros as mentioned in the introduction. The point made by this example is that technological innovations seems to always be preferred to social institutions when presented as solutions to different challenges irrespective, as it seems, of the calculated costs. Social institutions may sound unrealistic when offered as potential solutions, at least initially, but so are most often technological innovations too. The difference is the dominating ideology, which often seems to be based on the notion that any technology is possible as a solution to a problem. The obstacle is not lack of realism, usually reformulated as lack of vision,

DOI: 10.1057/9781137547125.0017

but adequate funding. Compare this to the general stance towards new bold social institutions. Here, calculations of adequate funding are seldom even discussed before ideas are dismissed as unrealistic, if not impossible.

What we thus need is a better balance between sanguine, techno-optimistic promises and one-dimensional views on accompanying institutional settings. The problem of how to manage spent nuclear fuel on a global scale shows that bold institutional solutions developed to exploit already existing technologies can be just as efficient, or at least just as interesting, as new technologies developed in existing institutional frameworks. As a conclusion, efforts and resources need not be spent only on developing new technologies, but also on how to develop institutions other than traditional market mechanisms, to make the world a better place for all its inhabitants.

In order to start to discuss alternatives to innovations as solutions to different problems, the existing variation of theoretical approaches to the relations between technology and institution needs to be reviewed. That was the supposition behind this book. Thus at the core of this study lies the multitude of perspectives developed in different contexts. To be sure, many of them, if not all, have components that have survived the contexts they were originally developed in. Marxism has been developed since Marx and keeps on supplying questions about class struggle in knowledge society just as the field of innovation studies is developed, for instance, to frame diverse institutional developments in concepts such as social innovation. What this incomplete review has shown, however, is that there are a number of different perspectives, developed in at least as many contexts and beyond these, general enough to still supply insights regarding the present situation. It is my firm belief that many of them would be valuable to expand further also in the context of innovation studies, especially if the field wants to maintain credibility in the claims to supply policy advice.

It is unquestionably to the disadvantage of institutional and technical development to limit policy advice to those narrow contexts most often supplied by innovation studies. And it is consequentially desirable to broaden the points of departure when giving counsel about how to make priorities between different technological and institutional alternatives. Because, as an old and often rehearsed declaration of innovation studies claims, it is the multitude of approaches that constitute the foundation of novel ideas. Policies of technology, institution and change would thus doubtlessly benefit from a wider approach than presently supplied by the core of innovation studies.

Despite their origin in historical relations between institution, technology and change, many of the theoretical frameworks reviewed here have a potential to, in varying degrees, change and improve today's situation. Some perspectives certainly have this as the ultimate goal, thus performativity. If it is indeed correct, as has often been assumed in different theoretical frameworks reviewed here, that the relations between institutions and technologies are growing closer and are even being dissolved, for example, reflected in analytical concepts such as hybridity, then policymakers, experts and consumers can be expected to get a stronger influence on technical change. Likewise, social researchers and their theories of relations between institution, technology and change can anticipate greater weight outside of academia. Against this backdrop, it is all the more important that theories express the heterogeneity of the relations they model and reason about.

Taken together, theories of institution, technology and change are caught between their contemporary relations from which they are formulated and their potential to be used for improvements and change. If accepted and applied, they have a tendency to both conserve and be applied to perform change. Thus there seems to be an ever more positive feedback loop between existing relations, the study of relations, the forming of relations and back to the existing relations again. This impact of analytical work on real-life experience, the interaction between the study of relations and the forming of relations has already been touched upon in Chapter 5 under the heading of performativity. But a growing potential for performativity is not enough if the plurality of theories we have access to does not meet the bill when it comes to precisely capturing the decisive features of today's changing relations between institution and technology.

In order to achieve even more accurate theories regarding our contemporary technological and institutional revolution, whether it has been going on since the 18th century or since yesterday, we need to further boost our creativity in multidisciplinary environments relying on both incisive critique and empirical data. The traditional recipe for invention and innovation, creativity and collaboration, is thus most likely also the remedy against notions of a struggle between nations for economic growth through technological innovation and instead the track to better theories regarding institutions and technologies in globalizing contexts.

First and foremost such theories need to take a broader scope than the traditional assumption of technology being developed by commercial business enterprises selling products or services on liberal markets competing with price, performance and a streamlined production process. We need to develop theories of technology, institution and change that depart from radically different assumptions in order to influence developments. My hope is only that this review can be a small step on that path.

Notes

1. An example of this perspective is: Manuel Castells, *The Information Age: Economy, Society and Culture*, 3 vols. (Oxford: Blackwell, 1996–1998).
2. The notion of a global repository has indeed been studied and suggested, see: Glenn Schweitzer & Kelly Robbins, eds, *Setting the Stage for International Spent Nuclear Fuel Storage Facilities: International Workshop Proceedings* (Washington, D.C: The National Academic Press, 2008).

Bibliography

Abel, Andrew B. & Janice C. Eberly (1996), "Optimal Investment with Costly Reversibility", *The Review of Economic Studies* 63:4, 581–593.

Abernathy, William J. & Kim B. Clark (1985), "Innovation: Mapping the Winds of Creative Destruction", *Research Policy* 14:1, 3–22.

Adler, Paul S. (2008), "Technological Determinism", in: *International Encyclopedia of Organization Studies*, eds, Stewart R. Clegg & James R. Bailey (London: SAGE Publications), 1537–1540.

Aghion, Philip & Peter Howitt (1998), *Endogenous Growth Theory* (Cambridge, Mass: The MIT Press).

—— (2012), *The Economics of Growth* (Cambridge, Mass: The MIT Press).

Akiike, Atsushi (2013), "Where Is Abernathy and Utterback Model?", *Annals of Business Administrative Science* 12, 225–236, accessed at: http://www.gbrc.jp/journal/abasjp/ms/abas12-17.pdf, March 18, 2014.

Akrich, Madeleine (1992), "The De-Scription of Technical Objects", in: *Shaping Technology/Building Society: Studies in Sociotechnical Change*, eds, Wiebe E. Bijker & John Law (Cambridge, Mass: The MIT Press), 205–224.

Alder, Ken (1997), *Engineering the Revolution: Arms and Enlightenment in France, 1763–1815* (Princeton: Princeton University Press).

Appadurai, Arjun (1986), "Introduction: Commodities and the Politics of Value", in: *The Social Life of Things: Commodities in Cultural Perspective*, ed., Arjun Appdaurai (Cambridge: Cambridge University Press), 3–63.

Bibliography

Arthur, W. Brian (2009), *The Nature of Technology: What It Is and How It Evolves* (New York: Free Press).

Aspers, Patrik (2007), "Theory, Reality, and Performativity in Markets", *The American Journal of Economics and Sociology* 66:2, 379–398.

Atzori, Luigi, Antonio Iera & Giacomo Morabito (2010), "The Internet of Things: A Survey", *Computer Networks* 54:15, 2787–2805.

Ayres, Clarence E. (1944), *The Theory of Economic Progress* (Chapel Hill: The University of North Carolina Press).

Badar, Karen (2003), "Posthumanist Performativity: Toward an Understanding of How Matter Comes to Matter", *Signs: Journal of Women in Culture and Society* 28:3, 801–831.

Baumol, William J. (2002), *The Free-Market Innovation Machine: Analyzing the Growth Miracle of Capitalism* (Princeton: Princeton University Press).

Beck, Ulrich (1992), *Risk Society: Towards a New Modernity*, original in German 1986 (London: SAGE Publications, 1992).

Bensaude-Vincent, Bernadette (2009), *Les vertiges de la technoscience: Façonner le monde atome par atome* (Paris: La Découverte).

Berman, Marshall (1988), *All That Is Solid Melts into Air: The Experience of Modernity*, original in 1982 (New York: Penguin Books).

Bijker, Wiebe E. (1995), *Of Bicycles, Bakelites, and Bulbs: Toward a Theory of Sociotechnical Change* (Cambridge, Mass: The MIT Press).

Bijker, Wiebe E. & John Law, eds (1992), *Shaping Technology/Building Society: Studies in Sociotechnical Change* (Cambridge, Mass: The MIT Press).

Bijker, Wiebe E., Thomas P. Hughes & Trevor Pinch, eds (1989), *The Social Construction of Technological Systems* (Cambridge, Mass: The MIT Press).

Bimber, Bruce (1994), "Three Faces of Technological Determinism", in: *Does Technology Drive History? The Dilemma of Technological Determinism*, eds, Merrit Roe Smith & Leo Marx (Cambridge, Mass: The MIT Press), 79–100.

Block Fred & Matthew R. Keller, eds (2010), *State of Innovation: The U.S. Government's Role in Technology Development* (Boulder, Co: Paradigm Publishers).

Boserup, Ester (1965), *The Conditions of Agricultural Growth: The Economics of Agrarian Change under Population Pressure* (London: Allen & Unwin).

Braun, Hans-Joachim (1980), "Gas oder Elektrizität? Zur Konkurrenz zweier Beleuchtungssysteme, 1880–1914", *Technikgeschichte* 47, 1–19.

Braverman, Harry (1974), *Labor and Monopoly Capital: The Degradation of Work in the Twentieth Century* (New York: Monthly Review Press).
Bright, Jr., Arthur A. (1949), *The Electric-Lamp Industry: Technological Change and Economic Development from 1800 to 1947* (New York: Macmillan).
Brooks, Harvey (1980), "Technology, Evolution and Purpose", *Dædalus* 109:1 (Winter), 65–81.
Bury, John Bagnell (1920), *The Idea of Progress: An Inquiry into Its Origin and Growth* (London: Macmillan).
Butler, Judith (1993), *Bodies that Matter: On the Discursive Limits of Sex* (London: Routledge).
Bynum, Terrell Ward (2004), "Ethical Challenges to Citizens of 'The Automatic Age': Norbert Wiener of the Information Society", *Journal of Information, Communication and Ethics in Society* 3:2, 65–74.
—— (2008), "Norbert Wiener and the Rise of Information Ethics", in: *Information Technology and Moral Philosophy*, eds, Jeroen van den Hoven & John Weckert (Cambridge: Cambridge University Press), 8–25.
—— (2010), "Philosophy in the Information Age", *Metaphilosophy* 41:3, 420–442.
Callon, Michel (1998), "An Essay on Framing and Overflowing: Economic Externalities Revisited by Sociology", in: *Laws of the Markets*, ed., Michel Callon (Oxford: Blackwell), 244–269.
—— (2005), "Why Virtualism Paves the Way to Political Impotence", *Economic Sociology: European Electronic Newsletter* 6:2, 3–20, accessed at: http://econsoc.mpifg.de/archive/esfeb05.pdf, February 20, 2014.
—— (2007), "What Does It Mean to Say that Economics Is Performative?", in: *Do Economists Make Markets? On the Performativity of Economics*, eds, Donald MacKenzie, Fabian Muniesa & Lucia Siu (Princeton: Princeton University Press), 311–357.
Carpenter Stephen R. et al. (2009), "Accelerate Synthesis in Ecology and Environmental Sciences", *BioScience* 59:8, 699–701.
Carrier, James G. & Daniel Miller (1998), *Virtualism: A New Political Economy* (Oxford: Berg).
Carter, Lemuria et al. (2012), "E-Government Utilization: Understanding the Impact of Reputation and Risk", *International Journal of Electronic Government Research* 8:1, 83–97.
Castells, Manuel (1996-1998), *The Information Age: Economy, Society and Culture*, 3 vols. (Oxford: Blackwell).

Cavalli-Sforza, Luigi Luca (1999), *Genes, Peoples, and Languages* (New York: North Point Press), 179–187.
Chandler, Alfred J. (1977), *The Visible Hand: The Managerial Revolution in American Business* (Cambridge, Mass: Harvard University Press).
—— (1990), *Scale and Scope: The Dynamics of Industrial Capitalism* (Cambridge, Mass: Harvard University Press).
Clarke, Robin (1974), "The New Utopias", *New Scientist* 62:898 (May 16), 423.
Coleman, Michael & David H. Gray (2014), "The Legality of Targeted Killings and the Use of Drones in the War on Terror", *Global Security Studies* 5:1, 37–55.
Constant II, Edward W. (1980), *The Origins of the Turbojet Revolution* (Baltimore: Johns Hopkins University Press).
Cottrill, Charlotte A., Everett M. Rogers & Tamsy Mills (1989), "Co-citation Analysis of the Scientific Literature of Innovation Research Traditions: Diffusion of Innovations and Technology Transfer", *Science Communication* 11:2, 181–208.
Courvisanos, Jerry (2012), *Cycles, Crises and Innovation: Path to Sustainable Development – A Kaleckian-Schumpeterian Synthesis* (Cheltenham: Edward Elgar).
Cowan, Ruth Schwartz (1989), "The Consumption Junction: A Proposal for Research Strategies in the Sociology of Technology", in: *The Social Construction of Technological Systems*, eds, Wiebe E. Bijker, Thomas P. Hughes & Trevor Pinch (Cambridge, Mass: The MIT Press), 261–280.
Cronon, William (1991), *Nature's Metropolis: Chicago and the Great West* (New York: W.W. Norton & Co).
Crouch, Tom D. (1992), "Why Wilbur and Orville? Some Thoughts on the Wright Brothers and the Process of Invention", in: *Inventive Minds: Creativity in Technology*, eds, Robert J. Weber & David N. Perkins (Oxford: Oxford University Press), 80–92.
Dardot, Pierre & Christian Laval (2013), *The New Way of the World: On Neoliberal Society*, original in French 2009 (London: Verso).
David, Paul A. (1985), "Clio and the Economics of QWERTY", *The American Economic Review* 75:2 (May), 332–337.
Dickson, David (1974), *Alternative Technology and the Politics of Technical Change* (London: Fontana Books).
Dietz, Thomas, Nives Dolšak, Elinor Ostrom & Paul C. Stern (2002), "The Drama of the Commons", in: *The Drama of the Commons:*

Committee on the Human Dimensions of Global Change, eds, Elinor Ostrom et al. (Washington, D.C: The National Academic Press), 3–35.

DiMaggio, Paul J. & Walter W. Powell (1983), "The Iron Cage Revisited: Institutional Isomorphism and Collective Rationality in Organizational Fields", *American Sociological Review* 48:2, 147–160.

Disco, Nil & Eda Kranakis (2013), "Toward a Theory of Cosmopolitan Commons", in: *Cosmopolitan Commons: Sharing Resources and Risks across Borders*, eds, Nil Disco & Eda Kranakis (Cambridge, Mass: The MIT Press), 57–96.

Dodgson, Mark & David Gann (2010), *Innovation: A Very Short Introduction* (Oxford: Oxford University Press).

Doloreux, David (2002) "What We Should Know about Regional Systems of Innovation", *Technology in Society* 24:3, 242–263.

Drack, Manfred (2009), "Ludwig von Bertalanffy's Early System Approach", *Behavioral Science* 26:5, 563–572.

Edgerton, David (2004), "'The Linear Model' Did Not Exist: Reflections on the History and Historiography of Science and Research in Industry in the Twentieth Century", in: *The Science-Industry Nexus: History, Policy, Implications*, eds, Karl Grandin, Nina Wormbs & Sven Widmalm, Nobel Symposium 123 (Sagamore Beach, Mass: Science History Publications), 31–57.

—— (2006), *The Shock of the Old: Technology and Global History since 1900* (London: Profile Books).

—— (2007), "Creole Technologies and Global Histories: Rethinking How Things Travel in Space and Time", *Journal of History of Science and Technology* 1, accessed at: http://johost.eu/vol1_summer_2007/vol1_de.htm, February 17, 2014, p. 23.

—— (2010), "Innovation, Technology, or History? What Is the Historiography of Technology About?", *Technology and Culture* 51:3, 680–697.

Edquist, Charles, ed. (1997), *Systems of Innovation: Technologies, Institutions and Organizations* (London: Pinter Publishers).

Edquist, Charles & Olle Edqvist (1979), "Social Carriers of Techniques for Development", *Journal of Peace Research* 16:4, 313–331.

Edsall, Thomas B. (2012), "No More Industrial Revolutions?", *The New York Times*, October 15, accessed at: http://campaignstops.blogs.nytimes.com/2012/10/15/no-more-industrial-revolutions/?_php=true&_type=blogs&module=Search&mabReward=relbias%3Ar%2C%5B%22RI%3A5%22%2C%22RI%3A18%22%5D&_r=0, June 30, 2012.

Edwards, Paul N. (1996), *The Closed World: Computers and the Politics of Discourse in Cold War America* (Cambridge, Mass: The MIT Press).

Elam, Mark (1993), *Innovation as the Craft of Combination: Perspectives on Technology and Economy in the Spirit of Schumpeter*, Linköping Studies in Arts and Science 95 (Linköping: Linköping University).

—— (1999), "Living Dangerously with Bruno Latour in a Hybrid World", *Theory, Culture & Society* 16:4, 1–24.

Elias, Norbert (1995), "Technization and Civilization", *Theory, Culture & Society* 12:3, 7–42.

Elliott, Anthony (2002), "Beck's Sociology of Risk: A Critical Assessment", *Sociology* 36:2, 293–315.

Ellul, Jacques (1964), *The Technological Society*, original in French 1954 (New York: Vintage Books).

Elster, Jon (1983), *Explaining Technical Change: A Case Study in the Philosophy of Science* (Cambridge: Cambridge University Press).

Elzinga, Aant (1984), "Research, Bureaucracy and the Drift of Epistemic Criteria", in: *The University Research System: The Public Policies of the Home of Scientists*, eds, Björn Wittrock & Aant Elzinga, Studies in Higher Education in Sweden 5 (Stockholm: Almqvist & Wiksell), 191–220.

—— (1997), "The Science-Society Contract in Historical Transformation with Special Reference to 'Epistemic Drift'", *Social Science Information* 36:3, 411–445.

Etzkowitz, Henry (2008), *The Triple Helix: University-Industry-Government Innovation in Action* (London: Routledge).

Etzkowitz, Henry & Loet Leydesdorff, eds (1997), *Universities and the Global Knowledge Economy: A Triple Helix of University-Industry-Government Relations* (London: Pinter Publishers).

—— (2000), "The Dynamics of Innovation: From National Systems and 'Mode 2' to a Triple Helix of University-Industry-Government Relations", *Research Policy* 29:2, 109–123

Fagerberg, Jan, Ben R. Martin & Esben S. Andersen (2013), "Innovation Studies: Towards a New Agenda", in: *Innovation Studies: Evolution and Future Challenges*, eds, Jan Fagerberg, Ben R. Martin & Esben S. Andersen (Oxford: Oxford University Press), 1–17.

Farrell, Joseph & Garth Saloner (1985), "Standardization, Compatibility, and Innovation", *The RAND Journal of Economics* 16:1, 70–83.

Feenberg, Andrew (1991), *Critical Theory of Technology* (Oxford: Oxford University Press).

—— (2000), "From Essentialism to Constructivism: Philosophy of Technology at the Crossroads", in: *Technology and the Good Life?*, eds, Eric Higgs, Andrew Light & David Strong (Chicago: The University of Chicago Press), 294–315.

—— (2000), "Do We Need a Critical Theory of Technology? Reply to Tyler Veak", *Science, Technology, & Human Values* 25:2, 238–242.

—— (2004), *Heidegger and Marcuse: The Catastrophe and Redemption of History* (London: Routledge).

—— (2010), *Between Reason and Experience: Essays in Technology and Modernity* (Cambridge, Mass: The MIT Press).

Feldman, Maryann P. (1994), *The Geography of Innovation*, Economics of Science, Technology and Innovation 2 (Dordrecht: Kluwer).

Ferguson, Eugen S. (1992), *Engineering and the Mind's Eye* (Cambridge, Mass: The MIT Press).

Findlay, C. Scott & Charles J. Lumsden (1988), "The Creative Mind: Toward an Evolutionary Theory of Discovery and Innovation", *Journal of Social and Biological Systems* 11:1, 3–55.

Finke, Ronald A. (1995), "Creative Insight and Preinventive Forms", in: *The Nature of Insight*, eds, R. J. Sternberg & J. E. Davidson (Cambridge, Mass: The MIT Press), 255–280.

Fischer, Claude S. (1992), *America Calling: A Social History of the Telephone to 1940* (Berkeley, Calif: University of California Press).

Fligstein, Neil (2001), *The Architecture of Markets: The Economic Sociology of Twenty-First-Century Capitalist Societies* (Princeton: Princeton University Press).

Florax, Raymond (1992), *The University: A Regional Booster* (Aldershot: Avebury).

Florida, Richard (2002), *The Rise of the Creative Class* (New York: Basic Books).

—— (2005), *Cities and the Creative Class* (New York: Routledge).

Forman, Paul (2007), "The Primacy of Science in Modernity, of Technology in Postmodernity, and of Ideology in the History of Technology", *History and Technology* 23:1, 1–152.

Franz, Hans-Werner, Josef Hochgerner & Jürgen Howaldt (2012), "Challenge Social Innovation: An Introduction", in: *Challenge Social Innovation*, eds, Hans-Werner Franz, Josef Hochgerner & Jürgen Howaldt (Dordrecht: Springer), 1–16.

Freeman, Christopher (1982), *The Economics of the Industrial Innovation*, 2nd ed. (London: Pinter Publishers).

Fridlund, Mats (1999), *Den gemensamma utvecklingen: Staten, storföretagen och samarbetet kring den svenska elkrafttekniken* (Stockholm: Symposion).

Fryer, David Ross (2004), *The Intervention of the Other: Ethical Subjectivity in Levinas and Lacan* (New York: Other Press).

Fuenfschilling, Lea & Bernhard Truffer (2014), "The Structuration of Socio-technical Regimes – Conceptual Foundations from Institutional Theory", *Research Policy* 43:4, 772–791.

Fuller, Steve (2011), *Humanity 2.0: What It Means to Be Human Past, Present and Future* (Basingstoke: Palgrave Macmillan).

Funke, Joachim (2009), "On the Psychology of Creativity", in: *Milieus of Creativity an Interdisciplinary Approach to Spatiality of Creativity*, eds, P. Meusburger, J. Funke & E. Wunder, Knowledge and Space, Vol. 2 (Heidelberg: Springer Science + Business Media, 2009), 11–23.

Galis, Vasilis & Anders Hansson (2012), "Partisan Scholarship in Technoscientific Controversies: Reflections on Research Experience", *Science as Culture* 21:3, 335–364.

Galis, Vasilis & Francis Lee (2014), "A Sociology of Treason: The Construction of Weakness", *Science, Technology, & Human Values* 39:1, 154–179.

Geels, Frank W. (2004), "From Sectoral Systems of Innovation to Socio-technical systems: Insights about Dynamics and Change from Sociology and Institutional Theory", *Research Policy* 33:6–7, 897–920.

Gibbons, Michael et al. (1994), *The New Production of Knowledge: The Dynamics of Science and Research in Contemporary Societies* (London: SAGE Publications).

Giedon, Siegried (1948), *Mechanization Takes Command: A Contribution to Anonymous History* (Oxford: Oxford University Press).

Giele, Janet Z. & Glen H. Elder, Jr. (1998), "Life Course Research: Development of a Field", in: *Methods of Life Course Research: Qualitative and Quantitative Approaches*, eds, Janet Z. Giele & Glen H. Elder, Jr. (London: SAGE Publications), 5–27.

Gispen, Kees (1989), *New Profession, Old Order: Engineers and German Society, 1815–1914* (Cambridge: Cambridge University Press).

Godin, Benoît (2006), "The Linear Model of Innovation: The Historical Construction of an Analytical Framework", *Science, Technology, & Human Values* 31:6, 639–667.

—— (2012), "'Innovation Studies': The Invention of a Speciality", *Minerva: A Review of Science, Learning and Policy* 50:4, 397–421.

—— (2014), "'Innovation Studies': Staking the Claim for a New Disciplinary 'Tribe'", *Minerva* 52:4, 489–495.
Godin, Benoît & Joseph P. Lane (2013), "Pushes and Pulls: Hi(S)tory of the Demand Pull Model of Innovation", *Science, Technology, & Human Values* 38:5, 621–654.
Gordon, Robert J. (2012), "Is U.S. Economic Growth Over? Faltering Innovation Confronts the Six Headwinds", NBER Working Papers 18315 (Cambridge, Mass: National Bureau of Economic Research, August), accessed at: http://faculty-web.at.northwestern.edu/economics/gordon/Is%20US%20Economic%20Growth%20Over.pdf, June 30, 2014.
Green, Stephen G., Mark B. Gavin & Lynda Aiman-Smith (1995), "Assessing a Multidimensional Measure of Radical Technological Innovation", *IEEE Transactions on Engineering Management* 42:3, 203–214.
Gruber, Howard E. & S. N. Davis (1988), "Inching Our Way Up Mount Olympus: The Evolving-Systems Approach to Creative Thinking", in: *The Nature of Creativity: Contemporary Psychological Perspectives*, ed., R. J. Sternberg (Cambridge: Cambridge University Press), 243–270.
Guston, David H. et al. (2014), "Responsible Innovation: Motivations for a New Journal", *Journal of Responsible Innovation* 1:1, 1–8.
Habermas, Jürgen (1970), *Toward a Rational Society: Student Protest, Science and Politics*, original in German in 1968 and 1969 (Boston: Beacon Press).
—— (1973), *Theory and Practice*, original in German in 1963, 1966 and 1968 (Boston: Beacon Press).
Hacking, Ian (1999), *The Social Construction of What?* (Cambridge, Mass: Harvard University Press).
Hall, Peter A. (2010), "Historical Institutionalism in Rationalist and Sociological Perspective", in: *Explaining Institutional Change: Ambiguity, Agency, and Power*, eds, James Mahoney & Kathleen Thelen (Cambridge: Cambridge University Press), 204–223.
Hall, Peter A. & David Soskice (2001), "An Introduction to Varieties of Capitalism", in: *Varieties of Capitalism: The Institutional Foundations of Comparative Advantage*, eds, Peter A. Hall & David Soskice (Oxford: Oxford University Press), 1–68.
Hamilton, David (1981), "Ayres' *Theory of Economic Progress*: An Evaluation of Its Place in Economic Literature", *American Journal of Economics and Sociology* 40:4, 427–438.

Hannaway, Owen (1986), "Laboratory Design and the Aim of Science: Andreas Libavius versus Tycho Brahe", *Isis* 77:4, 585–610.

Haraway, Donna J. (1997), *Modest_Witness@Second_Millennium. FemaleMan©_Meets_OncoMouse™: Feminism and Technoscience* (London: Routledge).

Hård, Mikael & Andrew Jamison (2005), *Hubris and Hybrids: A Cultural History of Technology and Science* (New York: Routledge).

Harvey, David (1989), *The Condition of Postmodernity: An Enquiry into the Conditions of Cultural Change* (Cambridge, Mass: Blackwell).

—— (2014), *Seventeen Contradictions and the End of Capitalism* (London: Profile Books).

Harwood, Jonathan (2010), "Understanding Academic Drift: On the Institutional Dynamics of Higher Technical and Professional Education", *Minerva: A Review of Science, Learning and Policy* 48:4, 413–427.

Hayles, N. Katherine (1999), *How We Became Posthuman: Virtual Bodies in Cybernetics, Literature, and Informatics* (Chicago: The University of Chicago Press).

Hecht, Gabrielle (1998), *The Radiance of France: Nuclear Power and National Identity after World War II* (Cambridge, Mass: The MIT Press).

Heinze, Thomas, Philip Shapira, Juan D. Rogers & Jacqueline M. Senker (2009), "Organizational and Institutional Influences on Creativity in Scientific Research", *Research Policy* 38:4, 610–623.

Hess, David, Steve Breyman, Nancy Campbell & Brian Martin (2008), "Science, Technology, and Social Movements", in: *The Handbook of Science and Technology Studies: Third Edition*, eds, Edward J. Hackett et al. (Cambridge, Mass: The MIT Press), 473–498.

Holm, Paul, Arne Jarrick & Dominic Scott (2015), *Humanities World Report 2015* (Basingstoke: Palgrave Macmillan).

Horkheimer, Max & Theodor W. Adorno (1944), *Dialektik der Auklärung* (New York: Social Studies Association).

Hounshell, David A. (1995), "Hughesian History of Technology and Chandlerian Business History: Parallels, Departures and Critics", *History and Technology* 12:3, 205–224.

—— (2004), "Industrial Research: Commentary", in: *The Science-Industry Nexus: History, Policy, Implications*, eds, Karl Grandin, Nina Wormbs & Sven Widmalm, Nobel Symposium 123 (Sagamore Beach, Mass: Science History Publications), 59–65.

Howells, John (2005), *The Management of Innovation and Technology* (London: SAGE Publications).
Hughes, Thomas P. (1983), *Networks of Power: Electrification of Western Society, 1880–1930* (Baltimore: Johns Hopkins University Press).
—— (1992), "The Dynamics of Technological Change: Salients, Critical Problems, and Industrial Revolutions", in: *Technology and Enterprise in a Historical Perspective*, eds, Giovanni Dosi, Renato Gianetti & Pier Angelo Toninelli (Oxford: Oxford University Press), 97–118.
—— (1994), "Technological Momentum", in: *Does Technology Drive History? The Dilemma of Technological Determinism*, eds, Merrit Roe Smith & Leo Marx (Cambridge, Mass: The MIT Press), 101–114.
Illich, Ivan (1974), *Energy and Equity*, World Perspectives 49 (New York: Harper & Row).
Immergut, Ellen M. (1998), "The Theoretical Core of the New Institutionalism", *Politics & Society* 26:1, 5–34.
Innovation Union homepage, accessed at: http://ec.europa.eu/research/innovation-union/index_en.cfm, March 17, 2014.
Jasanoff, Sheila (2004), "The Idiom of Co-production", in: *States of Knowledge: The Co-production of Science and Social Order*, ed., Sheila Jasanoff (London: Routledge), 1–12.
Jasanoff, Sheila & Sang-Hyun Kim (2009), "Containing the Atom: Sociotechnical Imaginaries and Nuclear Power in the United States and South Korea", *Minerva* 47:2, 119–146.
——, eds (forthcoming August 2015), *Dreamscapes of Modernity: Sociotechnical Imaginaries and the Fabrication of Power* (Chicago: The University of Chicago Press).
Joerges, Bernward (1999), "Do Politics Have Artefacts?", *Social Studies of Science* 29:3, 411–431.
—— (1999), "Scams Cannot Be Busted: Reply to Woolgar & Cooper", *Social Studies of Science* 29:3, 450–457.
Kahn, David (1991), *Seizing the Enigma: The Race to Break the German U-Boats Codes, 1939–1943* (Boston: Houghton Mifflin).
Kaiserfeld, Thomas (2013), "Why New Hybrid Organizations Are Formed: Historical Perspectives on Epistemic and Academic Drift", *Minerva: A Review of Science, Learning and Policy* 51:2, 171–194.
Kaku, Michio (2011), *The Physics of the Future: How Science Will Shape Human Destiny and Our Daily Lives by the Year 2100* (New York: Doubleday).

Karns Alexander, Jennifer (2012), "Thinking Again about Science in Technology", *Isis* 103:3, 518–526.

Katzenstein, Peter J. (1985), *Small Countries in World Markets: Industrial Policy in Europe* (Ithaca, NY: Cornell University Press).

Katznelson, Ira (2003), "Periodization and Preferences: Reflections on Purposive Action in Comparative Historical Social Science", in: *Comparative Historical Analysis in the Social Sciences*, eds, James Mahoney & Dietrich Rueschemeyerd (Cambridge: Cambridge University Press), 270–303.

Keenan, Michael (2003), "Identifying Emerging Generic Technologies at the National Level: The UK Experience", *Journal of Forecasting* 22:2–3, 129–160.

Kingston, Christopher & Gonzalo Caballero (2009), "Comparing Theories of Institutional Change", *Journal of Institutional Economics* 5:2, 151–180.

Kline, Ronald R. (2006), "Cybernetics, Management Science and Technology Policy: The Emergence of Information Technology as a Keyword, 1948–1985", *Technology and Culture* 47:3, 513–535.

Kranakis, Eda (2013), "The 'Good Miracle': Building a European Airspace Commons, 1919–1939", in: *Cosmopolitan Commons: Sharing Resources and Risks across Borders*, eds, Nil Disco & Eda Kranakis (Cambridge, Mass: The MIT Press), 13–55.

Kroeber, A. L. (1963), *Anthropology: Culture Patterns and Processes* (New York: Harcourt, Brace & World).

Kuhn, Thomas S. (1962), *The Structure of Scientific Revolutions* (Chicago: The University of Chicago Press).

Kurzweil, Raymond (1999), *The Age of Spiritual Machines* (New York: Viking Press).

—— (2005), *The Singularity Is Near: When Humans Transcend Biology* (New York: Penguin).

Langlois, Richard N. (2003), "Schumpeter and the Obsolescence of the Entrepreneur", in: *Austrian Economics and Entrepreneurial Studies*, eds, Roger Koppl, Jack Birner & Peter Kurrild-Klitgaard, Advances in Austrian Economics 6 (Bingley: Emerald Publishing), 283–298.

Latour, Bruno (1993), *We Have Never Been Modern*, original in French 1991 (Cambridge, Mass: Harvard University Press).

—— (2005), *Reassembling the Social: An Introduction to Actor-Network Theory* (Oxford: Oxford University Press).

Latour, Bruno & Steve Woolgar (1979), *Laboratory Life: The Construction of Scientific Facts* (London: SAGE Publications).
Law, John (1999), "After ANT: Complexity, Naming and Topology", in: *Actor Network Theory and After*, eds, John Law & John Hassard (Oxford: Blackwell), 1–14.
Layton, Edwin T. (1971), *The Revolt of the Engineers: Social Responsibility and the American Engineering Profession* (Cleveland: Press of Case Western Reserve University).
—— (1974), "Technology as Knowledge", *Technology and Culture* 15:1, 31–41.
Lecours, André (2005), "Introduction", in: *New Institutionalism: Theory and Practice*, ed., André Lecours (Toronto: University of Toronto Press), 3–25.
Liebowitz, S. J. & Stephen E. Margolis (1994), "Network Externality: An Uncommon Tragedy", *Journal of Economic Perspectives* 8:2, 133–150.
—— (1995), "Path Dependence, Lock-In, and History", *The Journal of Law, Economics, & Organization* 11:1, 205–226.
Lindgren, Håkan, ed. (1996), *Economic Dynamism*, The Institute for Economic and Business History Research, Research Report No. 6 (Stockholm: Stockholm School of Economics).
Lindgren, Michael (1987), *Glory and Failure: The Difference Engines of Johann Müller, Charles Babbage and Georg and Edvard Scheutz*, Linköping Studies in Arts and Science 9, Stockholm Papers in History and Philosophy of Technology 2017 (Linköping: Linköping University).
Lohmann, Larry (2008), "Carbon Trading, Climate Justice and the Production of Ignorance: Ten Examples", *Development* 51:3, 359–365.
—— (2010), "Neoliberalism and the Calculable World: The Rise of Carbon Trading", in: *The Rise and Fall of Neoliberalism: The Collapse of an Economic Order?*, eds, Kean Birch & Vlad Myhnenko (London: Zed Books), 77–93.
Longino, Helen (2005), "Whither Philosophy of Science?", *Studies in History and Philosophy of Science* 36:4, 774–778.
Lubart, Todd I. (2000–2001), "Models of the Creative Process: Past, Present and Future", *Creative Research Journal* 13:3–4, 295–308.
Lundvall, Bengt-Åke, ed. (1992), *National Systems of Innovation: Towards a Theory of Innovation and Interactive Learning* (London: Anthem Press).
Lyotard, Jean-François (1984), *The Postmodern Condition: A Report on Knowledge*, Theory and History of Literature, Vol. 10, original in French 1979 (Manchester: Manchester University Press).

Macfarlane, Alan & Martin Gerry (2002), *Glass: A World History* (Chicago: The University of Chicago Press).

Mack, Chris A. (2011), "Fifty Years of Moore's Law", *IEEE Transactions on Semiconductor Manufacturing* 24:2 (May), 202–207.

Mahoney, James (2000), "Path Dependence in Historical Sociology", *Theory and Society* 29:4, 507–548.

Mahoney, James & Kathleen Thelen (2010), "A Theory of Gradual Institutional Change", in: *Explaining Institutional Change: Ambiguity, Agency, and Power*, eds, James Mahoney & Kathleen Thelen (Cambridge: Cambridge University Press), 1–37.

Maine, Elicia & Elizabeth Garnsey (2006), "Commercializing Generic Technology: The Case of Advanced Materials Ventures", *Research Policy* 35:3, 375–393.

Maneschi, Andrea (2002), "The Tercentenary of Henry Martyn's *Considerations upon the East-India Trade*", *Journal of the History of Economic Thought* 24:2, 233–249.

Marcuse, Herbert (1964), *One-Dimensional Man: Studies in the Ideology of Advanced Industrial Society* (Boston: Beacon Press).

Markard, Jochen, Rob Raven & Bernhard Truffer (2012), "Sustainability Transitions: An Emerging Field of Research and Its Prospects", *Research Policy* 41:6, 955–967.

Marshall, Alfred (1919), *Industry and Trade* (London: Macmillan).

Martin, Ben R. (2013), "Innovation Studies: An Emerging Agenda", in: *Innovation Studies: Evolution and Future Challenges*, eds, Jan Fagerberg, Ben R. Martin & Esben S. Andersen (Oxford: Oxford University Press), 168–186.

Martin, Brian (1995), "Technological Determinism Revisited", *Metascience* 4:2, 158–160.

Martyn, Henry (1701), *Considerations upon the East-India Trade* (London).

Marx, Karl (1937), *The Poverty of Philosophy*, original in French in 1847 (London: Martin Lawrence).

Marx, Leo (2010), "Technology: The Emergence of a Hazardous Concept", *Technology and Culture* 51:3, 561–577.

Marx, Leo & Merrit Roe Smith (1994), "Introduction", in: *Does Technology Drive History? The Dilemma of Technological Determinism*, eds, Merrit Roe Smith & Leo Marx (Cambridge, Mass: The MIT Press), ix–xv.

Mattsson, Helena & Sven-Olov Wallenstein (2010), "Introduction", in: *Swedish Modernism: Architecture, Consumption and the Welfare State*, eds, Helena Mattsson & Sven-Olov Wallenstein (London: Black Dog Publishing), 6–33.

Mazzucato, Mariana (2013), *The Entrepreneurial State: Debunking Public vs. Private Sector Myths* (London: Anthem Press).

McKelvey, Maureen (1996), *Evolutionary Innovations: The Business of Biotechnology* (Oxford: Oxford University Press).

McNeill, William H. (1963), *The Rise of the West: A History of the Human Community* (Chicago: The University of Chicago Press).

Meyer, Herbert W. (1972), *A History of Electricity and Magnetism*, Burndy Library Publication No. 27 (Norwalk, Conn: Burndy Library).

Miller, Daniel (2002), "Turning Callon the Right Way Up", *Economy and Society* 31:2, 218–233.

——— (2005), "Reply to Michel Callon", *Economic Sociology: European Electronic Newsletter* 6:3, 3–13, accessed at: http://econsoc.mpifg.de/archive/esjuly05.pdf, February 20, 2014.

Miller, Seumas (2012), "Social Institutions", in: *The Stanford Encyclopedia of Philosophy* (Fall 2012 Edition), ed., Edward N. Zalta, accessed at: http://plato.stanford.edu/archives/fall2012/entries/social-institutions/, January 17, 2014.

Mirowski, Philip (2004), "The Scientific Dimensions of Social Knowledge and Their Distant Echoes in 20th-Century American Philosophy of Science", *Studies in History and Philosophy of Science* 35:2, 283–326.

——— (2005), "Hoedown at the OK Corral: More Reflections on the 'Social' in Current Philosophy of Science", *Studies in History and Philosophy of Science* 36:4, 790–800.

Mirowski, Philip & Edward Nik-Khah (2008), "Command Performance: Exploring What STS Thinks It Takes to Build a Market", in: *Living in a Material World: Economic Sociology Meets Science and Technology Studies*, eds, Trevor Pinch & Richard Swedberg (Cambridge, Mass: The MIT Press), 89–128.

Misa, Thomas J. (1994), "Retrieving Sociotechnical Change from Technological Determinism", in: *Does Technology Drive History? The Dilemma of Technological Determinism*, eds, Merrit Roe Smith & Leo Marx (Cambridge, Mass: The MIT Press), 115–141.

Misa, Thomas J., Philip Brey & Andrew Feenberg, eds (2003), *Modernity and Technology* (Cambridge, Mass: The MIT Press).

Mishan, Edward J. (1969), *Growth: The Price We Pay* (London: Staples Press).

Moser, Ingunn (2000), "Against Normalisation: Subverting Norms of Ability and Disability", *Science as Culture* 9:2, 201–240.

Mulgan, Geoff (2012), "Social Innovation Theories: Can Theory Catch Up with Practice?", in: *Challenge Social Innovation*, eds, Hans-Werner Franz, Josef Hochgerner & Jürgen Howaldt (Dordrecht: Springer), 19–42.

Mumford, Lewis (1934), *Technics and Civilization* (New York: Harcourt, Brace & Company).

Mumford, Michael D. (2002), "Social Innovation: Ten Cases from Benjamin Franklin", *Creativity Research Journal* 14:2, 253–266.

Munthe, Christian (2013), "A New Ethical Landscape of Prenatal Testing: Individualising Choice to Serve Autonomy and Promote Public Health. A Radical Proposal", Presentation at: Individualized Choice: A New Approach to Reproductive Autonomy in Prenatal Screening, Brocher Foundation, Geneva, April 4–5, 2013.

Murmann Johann Peter & Koen Frenken (2006), "Toward a Systematic Framework for Research on Dominant Designs, Technological Innovations, and Industrial Change", *Research Policy* 35:7, 925–952.

Muro, Mark & Bruce Katz (2011), "The New 'Cluster Moment': How Regional Innovation Clusters Can Foster the Next Economy", in: *Entrepreneurship and Global Competitiveness in Regional Economies: Determinants and Policy Implications*, eds, Gary D. Libecap & Sherry Hoskinson, Advances in the Study of Entrepreneurship, Innovation & Economic Growth 22 (Bingley: Emerald Publishing), 93–140.

Nederveen Pieterse, Jan (2012), "Cultural Hybridity", in: *Encyclopedia of Global Studies*, eds, Helmut K. Anheler & Mark Juergensmeyer, Vol. 1 (London: SAGE Publications).

Nelson, Richard R. (1995), "Recent Evolutionary Theorizing about Economic Change", *Journal of Economic Literature* 33:1 (March), 48–90.

―――― (2005), *Technology, Institutions, and Economic Growth* (Cambridge, Mass: Harvard University Press).

Nelson, Richard R. & Howard Pack (1999), "The Asian Miracle and Modern Growth Theory", *The Economic Journal* 109:457 (July), 416–436.

Nelson, Richard R. & Bhaven N. Sampat (2001), "Making Sense of Institutions as a Factor Shaping Economic Performance", *Journal of Economic Behaviour & Organization* 44:1, 31–54.

Nickles, Thomas (2014), "Scientific Revolutions", in: *The Stanford Encyclopedia of Philosophy* (Summer 2014 Edition), ed., Edward N.

Zalta, accessed at: http://plato.stanford.edu/archives/sum2014/entries/scientific-revolutions/, February 14, 2015.

Noble, David F. (1977), *America by Design: Science, Technology, and the Rise of Corporate Capitalism* (Oxford: Oxford University Press).

——— (1984), *Forces of Production: A Social History of Industrial Automation* (New York: Random House).

Norman, Donald A. (1993), *Things that Make Us Smart: Defending Human Attributes in the Age of the Machine* (Boston: Addison-Wesley).

North, Douglass (1990), *Institutions, Institutional Change and Economic Performance* (Cambridge: Cambridge University Press, 1990).

November, Joseph (2012), *Biomedical Computing: Digitizing Life in the United States* (Baltimore: Johns Hopkins University Press).

Nowotny, Helga (2006), "Introduction: The Quest for Innovation and Cultures of Technology", in: *Cultures of Technology and the Quest for Innovation*, ed., Helga Nowotny (New York: Berghahn Books), 1–23.

Nowotny, Helga, Peter Scott & Michael Gibbons (2001), *Rethinking Science: Knowledge and the Public in an Age of Uncertainty* (Cambridge: Polity Press).

——— (2003), "Introduction: 'Mode 2' Revisited: The New Production of Knowledge", *Minerva: A Review of Science, Learning and Policy* 41:3, 179–194.

O'Brien, Patrick K. (2009), "The Needham Question Updated: A Historiographical Survey and Elaboration", *History of Technology* 29, 7–28.

Oldenziel, Ruth, Adri A. Albert de la Bruhèze & Onno de Wit (2005), "Europe's Mediation Junction: Technology and Consumer Society in the 20th Century", *History and Technology* 21:1, 107–139.

Ortega, José Guadalupe (2014), "Machines, Modernity, and Sugar: The Greater Caribbean in a Global Context, 1812–50", *Journal of Global History* 9:1, 1–25.

Ostrom, Ellinor (2005), *Understanding Institutional Diversity* (Princeton: Princeton University Press).

Oudshoorn, Nelly & Trevor Pinch, eds (2005), *How Users Matter: The Co-construction of Users and Technology* (Cambridge, Mass: The MIT Press).

Pacey, Arnold (1990), *Technology in World Civilization: A Thousand-Year History* (Cambridge, Mass: The MIT Press).

Passer, Harold C. (1952), "Development of Large-Scale Organization: Electrical Manufacturing around 1900", *The Journal of Economic History* 12:4, 378–395.

Pavitt, Keith (1997), "Academic Research, Technical Change and Government Policy", in: *Science in the Twentieth Century*, eds, John Krige & Dominique Pestre (Amsterdam: Harwood Academic Publishers), 143–158.

Pelto, Pertti J. (1973), *The Snowmobile Revolution: Technology and Social Change in the Arctic* (Menlo Park, Calif: Cummings Publishing).

Perkins, David N. (1988), "The Possibility of Invention", in: *The Nature of Creativity: Contemporary Psychological Perspectives*, ed., R. J. Sternberg (Cambridge: Cambridge University Press), 362–385.

Perren, Lew & Jonathan Sapsed (2013), "Innovation as Politics: The Rise and Reshaping of Innovation in UK Parliamentary Discourse 1960–2005", *Research Policy* 42:10, 1815–1828.

Perrin, Noel (1979), *Giving Up the Gun: Japan's Reversion to the Sword, 1543–1879* (Boston: Godin).

Peters, B. Guy (2005), *Institutional Theory of Political Science: The 'New Institutionalism'*, 2nd ed. (London: Continuum).

Phills Jr., James A., Kriss Deiglmeier & Dale T. Miller (2008), "Rediscovering Social Innovation", *Stanford Social Innovation Review* 6:4, 34–43

Pinch, Trevor & Richard Swedberg (2008), "Introduction", in: *Living in a Material World: Economic Sociology Meets Science and Technology Studies*, eds, Trevor Pinch & Richard Swedberg (Cambridge, Mass: The MIT Press), 1–26.

Preda, Alex (2008), "STS and Social Studies of Finance", in: *The Handbook of Science and Technology Studies: Third Edition*, eds, Edward J. Hackett et al. (Cambridge, Mass: The MIT Press), 901–920.

Pritchard, Sara B. (2011), *Confluence: The Nature of Technology and the Remaking of the Rhône* (Cambridge, Mass: Harvard University Press).

Radkau, Jonathan (2008), *Nature and Power: A Global History of the Environment*, original in German 2002 (Cambridge: Cambridge University Press).

Ramke, Jacqueline, Renee Du Toit & Garry Brian (2006), "An Assessment of Recycled Spectacles Donated to a Developing Country", *Clinical & Experimental Ophthalmology* 34:7, 671–676.

Roberts, Dorothy & Sujatha Jesudason (2013), "Movement Intersectionality: The Case of Race, Gender, Disability, and Genetic Technologies", *Du Bois Review* 10:2, 313–328.

Rogers, Everett M. (1962), *Diffusion of Innovations* (New York: The Free Press).

—— (1976), "Where Are We in Understanding the Diffusion of Innovations?", in: *Communication and Change: The Last Ten Years– and the Next*, eds, Wilbur Schramm & Daniel Lerner (Honolulu: The University Press of Hawaii), 204–222.

Romer, Paul M. (1990), "Endogenous Technological Change", *The Journal of Political Economy* 98:5 (October), S71–S102.

Rosa, Hartmut (2013), *Social Acceleration: A New Theory of Modernity*, original in German 2005 (New York: Columbia University Press, 2013).

Rose, Nikolas (2006), *Politics of Life Itself: Biomedicine, Power and Subjectivity in the Twenty-First Century* (Princeton: Princeton University Press).

Rosenberg, Nathan (1974), "Science, Invention and Economic Growth", *The Economic Journal* 84:3 (March), 90–108.

—— (1976), *Perspectives on Technology* (Cambridge: Cambridge University Press, 1976), 202–206.

—— (1982), "Marx as a Student of Technology", in: *Inside the Black Box: Technology and Economics* (Cambridge: Cambridge University Press), 34–51.

—— (1994), *Exploring the Black Box: Technology, Economics, and History* (Cambridge: Cambridge University Press).

Rothstein, Bo (1991), "State Structure and Variations in Corporatism: The Swedish Case", *Scandinavian Political Studies* 14:2, 149–171.

—— (1992), *Den korporativa staten: Intresseorganisationer och statsförvaltning i svensk politik* (Stockholm: Norstedts).

Rowe, William & Vivian Schelling (1991), *Memory and Modernity: Popular Culture in Latin America* (London: Verso).

Samuels, Warren J. (1977), "Technology Vis-à-Vis Institutions in the JEI: A Suggested Interpretation", *Journal of Economic Issues* 11:4, 871–895.

Sawyer, P. H. & R. H. Hilton (1963), "Technical Determinism: The Stirrup and the Plough", *Past & Present* 24:1 (April), 90–100.

Schaller, Robert R. (1997), "Moore's Law: Past, Present, and Future", *IEEE Spectrum* 34:6 (June), 52–59.

Schatzberg, Eric (2006), "Technik Comes to America: Changing Meanings of Technology in America before 1930", *Technology and Culture* 47:3, 488–512.

—— (2012), "From Art to Applied Science", *Isis* 103:3, 555–563.

Schmookler, Jacob (1966), *Invention and Economic Growth* (Cambridge, Mass: Harvard University Press).

Schweitzer, Glenn & Kelly Robbins, eds (2008), *Setting the Stage for International Spent Nuclear Fuel Storage Facilities: International Workshop Proceedings* (Washington, D.C: The National Academic Press).

Sharp, Lauriston (1952), "Steel Axes for Stone-Age Australians", *Human Organization* 11:2, 17–22.

Shinn, Terry (2002), "The Triple Helix and New Production of Knowledge: Prepackaged Thinking on Science and Technology", *Social Studies of Science* 32:4, 599–614.

Sismondo, Sergio (1993), "Some Social Constructions", *Social Studies of Science* 23:3, 515–553.

—— (1996), *Science without Myth: On Constructions, Reality, and Social Knowledge* (Albany: State University of New York Press).

Slingerland, Edward (2008), *What Science Offers the Humanities: Integrating Body and Culture* (Cambridge: Cambridge University Press).

Smith, Adam (1776), *An Inquiry into the Nature and Causes of the Wealth of Nations*, 2 vols. (London).

Smith, Adrian (2005), "The Alternative Technology Movement: An Analysis of Its Framing and Negotiation of Technology Development", *Human Ecology Review* 12:2, 106–119.

Smith, Adrian & Andy Stirling (2010), "The Politics of Social-Ecological Resilience and Sustainable Socio-technical Transitions", *Ecology and Society* 15:1, accessed at: http://www.ecologyandsociety.org/vol15/iss1/art11/, May 20, 2014.

Smithers, Andrew (2013), *The Road to Recovery: How and Why Economic Policy Must Change* (Chichester: John Wiley & Sons).

Söderberg, Gabriel (2013), *Constructing Invisible Hands: Market Technocrats in Sweden 1880–2000*, Acta Universitatis Upsaliensis: Uppsala Studies in Economic History 98 (Uppsala: Uppsala University).

Sörlin, Sverker (2012), "Environmental Humanities: Why Should Biologists Interested in the Environment Take the Humanities Seriously?", *BioScience* 62:9, 788–789.

Sörlin, Sverker & Hebe Vessuri (2007), "Introduction: The Democratic Deficit of Knowledge Economies", in: *Knowledge Society vs. Knowledge Economy: Knowledge, Power, and Politics*, eds, Sverker Sörlin & Hebe Vessuri (Basingstoke: Palgrave Macmillan), 1–33.

Stahl, Roger (2010), *Militainment Inc. War, Media, and Popular Culture* (New York: Routledge).

Stapleton, Larry (2014), "Technology, Culture and International Stability", *AI & Society* 29:2, 139–142.

Star, Susan Leigh (1991), "Power, Technology and the Phenomenology of Conventions: On Being Allergic to Onions", in: *A Sociology of Monsters: Essays on Power, Technology and Domination*, ed., John Law, Sociological Review Monographs 38 (London: Routledge), 26–56.

——— (1992), "The Trojan Door: Organizations, Work, and the 'Open Black Box'", *Systems Practice* 5:4, 395–410.

Sterio, Milena (2012), "The United States' Use of Drones in the War on Terror: The (Il)legality of Targeted Killings under International Law", *Case Western Reserve Journal of International Law* 45:1–2, 197–214.

Streeck, Wolfgang (2009), "Institutions in History: Bringing Capitalism Back In", Max Planck Institute for the Study of Societies, Discussion Paper 09/8, November.

Streeck, Wolfgang & Kathleen Thelen (2005), "Introduction: Institutional Change in Advanced Political Economies", in: *Beyond Continuity: Institutional Change in Advanced Political Economies*, eds, Wolfgang Streeck & Kathleen Thelen (Oxford: Oxford University Press), 3–39.

Styhre, Alexander & Mats Sundgren (2005), *Managing Organization Creativity: Critique and Practices* (Basingstoke: Palgrave).

Sundqvist, Göran (2002), *The Bedrock of Opinion: Science, Technology and Society in the Siting of High-Level Nuclear Waste* (Dordrecht: Kluwer).

Suppes, Patrick (2001), "Weak and Strong Reversibility of Causal Processes", *Stochastic Causality*, eds, M. C. Galavotti, P. Suppes & D. Constantini (Stanford: CSLI Publications), 203–220.

Swidler, Ann (1986), "Culture in Action: Symbols and Strategies", *American Sociological Review* 51:2, 273–286.

Szetu, John et al. (2007), "Recycled Donated Spectacles: Experiences of Eye Care Personnel in the Pacific", *Clinical & Experimental Ophthalmology* 35:4, 391–392.

Tabachnick, David Edward (2007), "Heidegger's Essentialist Responses to the Challenge of Technology", *Canadian Journal of Political Science* 40:2, 487–505.

Tassey, Gregory (2007), *The Technology Imperative* (Cheltenham: Edward Elgar Publishing).

Tatarkiewicz, Wladyslaw (1980), *A History of Six Ideas: An Essay in Aesthetics*, Melbourne International Philosophy Series 5 (The Hague: Martinus Nijhoff Publishers).

Taylor, Peter (1995), "Co-construction and Process: A Response to Sismondo's Classification of Constructivisms", *Social Studies of Science* 25:2, 348–359.
Tolbert, Pamela S. & Lynne G. Zucker (1996), "The Institutionalization of Institutional Theory", in: *Handbook of Organization Studies*, eds, S. Clegg, C. Hardy & W. Nord (London: SAGE Publications), 175–190.
Totman, Conrad (1980), "Review", *Journal of Asian Studies* 39:3, 599–601.
Toynbee, Arnold J. (1934–1961), *A Study of Civilizations*, 12 vols. (Oxford: Oxford University Press).
Tydén, Mattias & Urban Lundberg (2010), "In Search of the Swedish Model: Contested Historiography", in: *Swedish Modernism: Architecture, Consumption and the Welfare State* eds, Helena Mattsson & Sven-Olov Wallenstein (London: Black Dog Publishing), 36–49.
Utterback, James M. & William J. Abernathy (1975), "A Dynamic Model of Product and Process Innovation", *Omega* 3:6, 639–656.
Varga, Attila (1998), *University Research and Regional Innovation: A Spatial Econometric Analysis of Academic Technology Transfer* (Dordrecht: Kluwer).
Vassilou, M. S. (2009), *Historical Dictionary of the Petroleum Industry*, Historical Dictionaries of Professions and Industries 4 (Plymouth: Scarecrow Press).
Veak, Tyler (2000), "Whose Technology? Whose Modernity? Questioning Feenberg's *Questioning Technology*", *Science, Technology, & Human Values* 25:2, 226–237.
Vergne, Jean-Philippe (2013), "QWERTY Is Dead; Long Live Path Dependency", *Research Policy* 42:6–7, 1191–1194.
Vermeulen, Niki, Sakari Tamminen & Andrew Webster, eds (2012), *Bio-Objects: Life in the 21st Century* (London: Ashgate).
Vincenti, Walter G. (1990), *What Engineers Know and How They Know It: Analytical Studies from Aeronautical History* (Baltimore: Johns Hopkins University Press).
Virilio, Paul (1999), *The Politics of the Very Worst*, ed., Sylvère Lotringer (Los Angeles: Semiotext(e)).
—— (2000), *The Information Bomb*, original in French 1998 (London: Verso).
—— (2006), *Speed and Politics*, original in French 1977 (Los Angeles: Semiotext(e)).

Waelbers, Katinka & Philipp Dorstewitz (2013), "Ethics in Actor Networks, or: What Latour Could Learn from Darwin and Dewey", *Science and Engineering Ethics* 20:1, 23–40.

Waldby, Catherine & Robert Mitchell (2006), *Tissue Economies: Blood, Organs and Cell Lines in Late Capitalism* (Durham, NC: Duke University Press).

Walker Rettberg, Jill (2014), *Seeing Ourselves through Technology: How We Use Selfies, Blogs and Wearable Devices to See and Shape Ourselves* (Basingstoke: Palgrave Macmillan).

Walsh, Vivien (1984), "Invention and Innovation in the Chemical Industry: Demand-Pull or Discovery-Push?", *Research Policy* 13:4, 211–234.

Weingart, Peter (1997), "From 'Finalization' to 'Mode 2': Old Wine in New Bottles?", *Social Science Information* 36, 591–613.

White Jr., Lynn (1962), *Medieval Technology and Social Change* (Oxford: Clarendon Press).

——— (1967), "The Historical Roots of Our Ecological Crisis", *Science* 155:3767 (March), 1203–1207.

Widmalm, Sven (2008), "History of Science in the Age of Policy", in: *Aurora Torealis: Studies in the History of Science and Ideas in Honor of Tore Frängsmyr*, eds, Marco Beretta, Karl Grandin & Svante Lindqvist (Sagamore Beach: Science History Publications), 259–275.

——— (2013), "Innovation and Control: Performative Research Policy in Sweden", in: *Transformations in Research, Higher Education and the Academic Market: The Breakdown of Scientific Thought*, eds, Sharon Rider, Ylva Hasselberg & Alexandra Waluszewski, Higher Education Dynamics 39 (Dordrecht: Springer, 2013), 39–51.

Winner, Langdon (1977), *Autonomous Technology: Technics-Out-of-Control as a Theme in Political Thought* (Cambridge, Mass: The MIT Press).

——— (1980), "Do Artifacts Have Politics?", *Daedalus: Journal of the American Academy of Arts and Sciences* 109:1, 121–136.

——— (1983), "Techne and Politeia: The Technical Constitution of Society", in: *Philosophy and Technology*, eds, Paul T. Durbin & Friedrich Rapp (Dordrecht: Kluwer), 97–111.

——— (1993), "Upon Opening the Black Box and Finding It Empty: Social Constructivism and the Philosophy of Technology", *Science, Technology, & Human Values* 18:3, 362–378.

——— (2005), "Resistance Is Futile: The Posthuman Condition and Its Advocates", in: *Is Human Nature Obsolete? Genetics, Bioengineering, and the Future of the Human Condition*, eds, Harold W. Baillie & Timothy K. Casey (Cambridge, Mass: The MIT Press), 385–410.

"Wireless power", *Wikipedia*, accessed at: http://en.wikipedia.org/wiki/Wireless_power#See_also, April 9, 2014.

Wise, George (1980), "A New Role for Professional Scientists in Industry: Industrial Research at General Electric, 1900–1916", *Technology and Culture* 21:3, 408–429.

——— (1983), "Ionists in Industry: Physical Chemistry at General Electric, 1900–1915", *Isis* 74:1, 7–21.

——— (1985), *Willis R. Whitney, General Electric, and the Origins of U.S. Industrial Research* (New York: Columbia University Press).

Woolgar, Steve & Geoff Cooper (1999), "Do Artefacts Have Ambivalence: Moses' Bridges, Winner's Bridges and Other Urban Legends in S&TS", *Social Studies of Science* 29:3, 433–449.

World Nuclear Association, accessed at: http://www.world-nuclear.org/info/nuclear-fuel-cycle/nuclear-wastes/radioactive-waste-management/, January 17, 2015.

———, accessed at: http://www.world-nuclear.org/info/nuclear-fuel-cycle/power-reactors/generation-iv-nuclear-reactors/, January 17, 2015.

Wormbs, Nina (2011), "Technology-Dependent Commons: The Example of Frequency Spectrum for Broadcasting in Europe in the 1920s", *International Journal of the Commons* 5:1, 92–109.

Wyatt, Sally (2008), "Technological Determinism Is Dead: Long Live Technological Determinism", in: *The Handbook of Science and Technology Studies: Third Edition*, eds, Edward J. Hackett et al. (Cambridge, Mass: The MIT Press), 165–180.

Zachmann, Karin (2002), "A Socialist Consumption Junction: Debating the Mechanization of Housework in East Germany, 1956–1957", *Technology & Culture* 43:1, 73–99.

Zalasiewicz, Jan et al. (2010), "The New World of the Anthropocene", *Environmental Science & Technology* 44:7, 2228–2231.

Zaloom, Caitlin (2006), *Out of the Pits: Traders and Technology from Chicago to London* (Chicago: Chicago University Press).

Ziman, John (2001), *Real Science: What It Is, and What It Means* (Cambridge: Cambridge University Press).

Index

acceleration, technological change, 114–117
action arenas, 32
activism, 42, 44, 59
actor-network theory, 60–63, 68
agency
   actor-network theory, 60–63
   entrepreneurship and, 58
   social construction, 58–59
   socio-technical regimes, 59–60
   technopolitical regimes, 59–60
automobile development, 71
autonomy, technology, 103, 114–115

Beck, Ulrich, 106–107
Berman, Marshal, 107–108
Bertalanffy, Ludwig von, 113
bioengineered-techno-body, 63, 120
biopolitics, 63
Boserup, Ester, 78
bottleneck, 72

change
   evolutionary economics, 38–41
   technical, 28–30, 33, 94, 108, 133
   technological and institutional, 6, 8, 13, 21, 39, 44, 80–81, 84–85, 91, 95–97, 99, 103, 106, 108, 113, 116, 121, 129
technology, institutions and, 20, 44, 53, 58, 68, 74, 78, 84, 95–96, 106, 129, 132–134
clock, 14, 97–98
co-construction, machine tools, 14–15
cognitive artefacts, 107–108
common-pool resources, 89–91
commons (common property resource management), 89–91
computers, 30, 112–113
conservative inventions, 21
constructivism, 105
   political, 106
   social, 59–60, 105–106
conversion, new institutionalism, 80–81
counterterrorism, Unmanned Aerial Vehicles (UAV), 12–13, 125
creativity
   human, 20, 123, 133
   models of process, 53
   technological frame, 122
creole technologies, 124
creolization, 125–126, 130
critical junctions, 81
critical theory, technology, 17
cultures, knowledge, 51–52

cybernetics, information, 112–113, 117, 129
cyborgs, 62, 112, 120

determinism, 103
  innovation, 8
  soft, 94
  strong, 94, 97
  technological, 94–95, 115
  weak, 94
development block, 70, 78
development pairs, 33
DiMaggio, Paul J., 82
Displacement, new institutionalism, 80
disruptive inventions, 21, 29
drift, new institutionalism, 80

economic growth
  education as indicator, 68
  entrepreneur as hero, 129
  innovation, 2–4, 8, 15, 42, 48, 131, 133
  motivation, 18
  technologies, 39, 95
economic performativity, 43–44
economy, free-market, 38
Edgerton, David, 122–124
education, 6, 28, 31, 33
  institution, 39, 51–52, 73, 82
  regional innovation activities, 68–69
electricity, 21, 60, 73, 98
Ellul, Jacques, 104, 108
endogenous factors
  change, 38
  institutional change, 39
  knowledge, 49
  research, 49, 51
  technological change, 39, 48, 50, 59, 89, 99
engineering, knowledge, 16–17
entrepreneur, 2, 8, 22, 39, 49, 58, 129
environment
  agricultural technologies, 78
  co-construction, 15
  creativity in, 122, 133
  economic, 38, 40, 50
  innovations, 2, 6
  modern, 107
  natural, 90, 98

institutional, 22, 63
sustainability, 13
technologies, 97, 124
urban, 12, 69
envirotechnical regimes, 89
eotechnics, 97–98
Euratom (European Atomic Energy Community), 5
European Commissioner for Research, Innovation and Science, 5
European Union, 2–3, 5, 33
European Union Emission Trading Scheme, 43, 129
evolution, science and technology, 39–40
evolutionary economics, 7, 38–40, 48
  Marxism and, 96, 99
  theories of, 38, 79
exhaustion, new institutionalism, 80–81
exogenous factors
  institutional change, 80–81, 99, 114
  inventions, 51, 58
  research, 49
  technological change, 18, 48, 50, 95–96, 99, 114

Feenberg, Andrew, 17
fractalization, space, 115–117
Fridlund, Mats, 33

Galbraith, John Kenneth, 103
gender, 42, 121
Generation IV reactors, nuclear power, 4–5, 63–64, 131–132
generic technologies, 99, 125
genetic technologies, 106, 125
glocal technologies, 126

Habermas, Jürgen, 106
Haraway, Donna, 62–63
Hård, Mikael, 120
Harvey, David, 114, 116
Heidegger, Martin, 105
historical institutionalism, 79
humanities
  debate on future of, 44–45
  global challenges, 6
hybridity, 120–122, 125, 130

hybridization
  humans, 61, 113–114
  technology and institution, 121–122, 125
hydropower, watermills, 89

incandescent light bulbs, 72–73
increasingly costly reversibility, 84–85
incremental inventions, 21
information system, 112–113
innovation, 2
  interaction between organizations and actors, 69–70
  linking technology and institution, 130–131
  monomania, 3, 130–131
  regional cluster, 68–69
  satisfying market demand, 48
innovation paradigm, 3, 22
innovation society, 3
innovation studies, 2–5, 7–8
  commercialization of new products, 129
  determinism, 99
  economics, 38, 131
  geographical aspects, 69
  ideal world of, 30
  integration and interaction of expertise, 122
  knowledge, 50, 58, 68, 132
  path dependence, 78
  performativity, 42
  policymaking, 48, 132–133
  resistance to change, 82, 89
Innovation Union, 2
institutional change, 13–14
  endogenous, 39
  exogenous, 80–81, 99, 114
  isomorphism, 81–82, 89, 130
  social transformations, 21–22
institutional imperative, 104
institutionalism
  historical, 79
  new, 31, 79–80, 82, 85, 89, 114
institutional lag, 12–13, 94
institutional logics, 60
institutions

innovation, 2, 7
  planning and designing, 7–8
  relations between technology and, 19–20, 129–130, 133–134
  social order, 18–19
  underdeveloped, 4–5, 63–64
international airspace, 91
International Prototype Kilogram, 81
internet of things, 62–63
inventions
  conservative (incremental), 21
  driving force, 48–49
  exogenous, 51, 58
  Marx's theory of, 96
  property rights, 31
  radical (disruptive), 21, 29
  technological frame, 52
  transformational, 31, 94, 99, 113
isomorphism, institutional change, 81–82, 89, 130

Jamison, Andrew, 120
Japan, 31, 39, 81, 123, 130
Judaeo-Christian tradition, 20, 90, 103

knowledge
  accumulation, 52–53
  cultural development, 51–52
  endogenous, 49
  engineering, 16–17
  inter- and intra-generational exchange, 51–52
  linear model, 49
  production and use, 50–51
  technology, 15–16, 52
knowledge society, 6, 49, 132
Kubrick, Stanley, 112

Latour, Bruno, 113–114, 120
layering, new institutionalism, 80
light bulbs, 72–73
linear model, knowledge, 49
Lyotard, Jean-François, 112–113

McKelvey, Maureen, 39
manufacturing, technology, 15–16

Marcuse, Herbert, 104–105, 108
market architecture, 31–32, 42
market institutions
  action arenas, 32
  diffusion of innovations, 29–30
  exchange between supplier and consumer, 28
  innovation by push of supply side, 29
  mediation junctions, 31–33
  private and public markets, 28–29
  technical change, 30–31
Marshall, Alfred, 68
Marx, Karl, 96–97, 99
Marxism, 44, 96–97, 99, 105–106, 121, 129, 132
mediation junctions
  development pairs, 33
  existence of, 32
  facilitation and creation, 31
migration, 52, 114, 122
militainment, 125
Misa, Thomas, 60, 95, 114
Mishan, Edward, 95
mobility, 52, 114, 122
modernity, 8, 20, 103–104, 106–108
  determinism through institutional change, 103
  institutional change, 106–108
  Latour as critic of, 113
  property rights, 31
  scientific knowledge, 42
  social acceleration characterizing, 115–117
  technological change, 103–104, 106–108
Moore's law, 99, 117
multiculturalism, 52, 114, 122
multiple invention, 72, 94
Mumford, Lewis, 97–99

nature, commons, 90
neocorporatism, 32–33
neotechnics, 98
new institutionalism, 31, 79–80, 82, 85, 89, 114

North, Douglass, 18
nuclear power
  demand for power, 123
  Generation IV reactors, 4–5, 63–64, 131–132
  spent fuel, 4–5, 63–64, 129, 132
nylon fibres, 120

paleotechnics, 98
patents, 31, 72–73
path dependence
  innovation studies, 78, 89
  new institutionalism, 82–85
performativity
  activism, 42, 44, 59
  economic, 43–44
  forming of relations, 133
  innovation, 7–8
  institutional theory, 44
  representationalism, 42
  virtualism, 43–44
pin factory, watches, 13–14
Polhem, Christopher, 14
political order, 105
political struggle, 105–106
posthumanism, 63, 112, 120–121
postmodernism, 8, 19, 63, 112–113, 117, 121, 129
Powell, Walter, 82
power
  hydropower, 89
  wind energy, 104
  see also nuclear power
prenatal testing, 114, 125
property rights, 31

QWERTY, path dependence, 82–83

radical inventions, 21, 29
Radkau, Jonathan, 124
regional cluster, innovation, 68–69
renewable power, wind energy, 104
representationalism, 42
research, 29, 38–39, 45, 52
  biotechnology, 63
  education, 68

research – *Continued*
  endogenous, 49, 51
  environmental, 15
  exogenous process, 49
  human, 20, 79–81
  industrial, 72
  military, 113
  nuclear power, 4–5
resistance to change, 78–85
reverse salients, 72
reversibility, increasingly costly, 84–85
risk society, 106
Rosa, Hartmut, 115–116
Rosenberg, Nathan, 50, 78–79

Sàmi, Skolt, 94
Schmookler, Jacob, 48–49
Schumpeter, Joseph, 2, 49, 58
science and technology, 6, 39, 42, 50, 53, 59, 106
secrecy, cipher and decipher code, 70–71
sedimentation, 80
simultaneous invention, 72, 94
Smith, Adam, 13
social constructivism, 59–60, 105–106
social innovations, 22
social institutions, 18
social order, 42
  institutions, 18–19
  technology, 105
social science
  debate on future of, 44–45
  global challenges, 6
  policymaking, 42
socio-technical regime, 59–60, 70, 78
socio-technical systems, 72–74, 78, 84
socio-technical transition, 21
soft determinism, 94
Sombart, Werner, 99
space compression, 114–116
Streeck, Wolfgang, 80
strong determinism, 94, 97

technical change, 28–30, 33, 94, 108, 133
technocracy, 106
technological change, 12–14
  acceleration, 114–117
  development blocks, 70
  endogenous, 39, 48, 50, 59, 89, 99
  exogenous, 18, 48, 50, 95–96, 99, 114
  inter- and intra-generational knowledge exchange, 51–52
  knowledge, 58
  performativity, 42–45
technological determinism, 94–95, 115
technological frame, 52, 62, 122
technological imperative, 104
technological momentum, 73, 78, 84
technological transfer, 8, 71, 123–125
technology
  autonomy, 103, 114–115
  exogenous, 18
  increasingly costly reversibility, 84–85
  manufacturing, 15–16
  nuclear power, 4–5, 63–64, 129, 132
  relations between institutions and, 19–20, 129–130, 133–134
  science and, 6, 39, 42, 50, 53, 59, 106
technopolitical regime, 59–60
Thelen, Kathleen, 80
transformational invention, 31, 94, 99, 113
transhumanism, 120–121
unity of disunity, 107–108

Unmanned Aerial Vehicles (UAV), counterterrorism, 12–13, 125
urban environment, 12, 69
urbanity, 103

video standards, VHS and Betamax, 83–84
Vienna declaration of 2011, 22
Virilio, Paul, 115–116
virtualism, performativity, 43–44

watch, clock manufacturing, 14
weak determinism, 94
Wealth of Nations (Smith), 13
White Jr., Lynn, 94

Wiener, Norbert, 113
wind energy, 104
Winner, Langdon, 103
Wormbs, Nina, 90

The manufacturer's authorised representative in the EU is Springer Nature Customer Service Centre GmbH, Europaplatz 3, 69115 Heidelberg, Germany. If you have any concerns regarding our products, please contact ProductSafety@springernature.com

Printed and bound by CPI Group (UK) Ltd, Croydon, CR0 4YY

23/03/2026

02076355-0020